GENERAL ARRANGEMENT.

ONS

RATOR

INEER
ATOR
GUNNER
ER
T GUNNER
R

REAR DECK

TAIL WALKWAY

R.P.B. ALLEN

13. DISTRESS FLARES
14. RECONNAISSANCE FLARES & CHUTE
15. CAMERA HATCH
16. FUEL TANKS

THE LAST
Flying Boat
ML814–ISLANDER

AROUND THE WORLD IN 50 YEARS

Half-title page : A sergeant flight engineer inspecting a Bristol Pegasus engine from the maintenance platforms.

(Photo Imperial War Museum, Ref CH 8561)

Above : A wartime view of Pembroke Dock, looking across the water to Neyland. Note the effective camouflage of the station's two large hangars, painted to resemble rows of terraced houses.

(Photo 422 Squadron Association)

THE LAST
Flying Boat
ML814–ISLANDER

AROUND THE WORLD IN 50 YEARS

PETER SMITH

Foreword by Captain Ken Emmott

Ensign
PUBLICATIONS

Mooring up. With the turret retracted and having picked up their mooring, crew members attach the storm pendant to make the Sunderland more secure in heavy weather.

(Photo Imperial War Museum, Ref CH 847)

© **Copyright Peter Smith 1993.**

First published in 1993 by
Ensign Publications
Hampshire Books Ltd.,
2 Redcar Street,
Southampton SO1 5LL
England. United Kingdom.

All rights reserved.

No part of this publication may be reproduced, stored in a retrieval system or transmitted in any form or by any means, electronic, mechanical, photocopying, recording or otherwise without the prior permission of Ensign Publications.

The right of Peter Smith to be identified as the author of this work has been asserted by him in accordance with the Copyright, Designs and Patents Act 1988.

Publisher David Graves.
Editorial consultant James Graves.
Designed and typeset by The Precinct Press.
Line illustrations by Robin Allen.
Cover by The Design Laboratory.
Cover photographs by François Prins.
Text repro. Alphaset, Southampton.
Photo. repro by MRM Graphics, Aylesbury.
Printed by Short Run Press, Exeter.

ISBN 185455 083 7

Contents

	Acknowledgments	6
	Foreword by Capt Ken Emmott	7
1	March 1944 and the Battle of the Atlantic	9
2	Into Service with 201 Squadron	17
3	Under Canadian Ownership	45
4	North and Further North	61
5	Pacific Days	73
6	Flying Boat Airlines	85
7	*Islander* and the Island	99
8	The Caribbean	119
9	Back in European Waters	141
10	Dockyard Days	165
11	Ireland and Beyond	193
	Appendices	214
	Glossary and Abbreviations	217
	Index	218
	Bibliography	224

Acknowledgments

A large number of people from all over the world have gone to considerable effort on my behalf, and quite simply without their assistance this book would not have been possible. With so many, all I can really attempt by way of thanks is to list them, although there are a few whom I feel that I must single out.

Firstly, the many individuals, commercial organisations and museums who have provided the photographs. These have all been provided without, or with a much reduced, reproduction fee, and without this assistance the book would not have been commercially viable. Those who have contributed photographs are all credited individually where the photographs appear in the book.

Secondly, François Prins, without whose constant advice and encouragement I doubt I would have ever seen the project through. Others who made outstanding contributions were: Erik Dokken of 330 Squadron, Royal Norwegian Air Force. John Laing, who provided almost all of the New Zealand information. Chris Murray and other staff of the Lord Howe Island Museum. Robin Allen, for the excellent line drawings and maps. Ian McIntyre, for all his hard work in producing the photographs.

I now list, hopefully without omissions, all those who have contributed to the project.

From the UK and Ireland: Robin Allen, Chaz Bowyer, Clive Brookes, Calshot Activities Centre staff, Denis Calvert, Nick Corrie, Ernie Cromie, Barry Eagles, Capt Dave Easton, Capt Ken Emmott, John Evans, Dick Froggatt, Willie Hamilton, Ron Harris, Andrew Hendrie, Capt Vic Hodgkinson, Edward Hulton, Ray Lassiter, Rev Peter Lillingston, Donald Lindsay, Ian McIntyre, Margaret O'Shaughnessy, Ron Parsons, Ron Richmond, Short Brothers PLC (Alec McRitchie, Michael Loader), Capt Keith Sissons, Angela Smith, John Stroud, Peter Thomas MBE, Sir Peter Thorne, KCVO, MBE, Capt John Vickers, Sqn Ldr Fred Weaver.

From Canada: Although 422 Squadron was disbanded at the end of the Second World War, its Squadron Association is still a very active. It was through this association that I was able to contact all those listed. Canadian Forces Photographic Unit (Ottawa), Norman Dodd, Jean Doern, Capt Gus Gauss, Larry Giles, Capt Jack Logan, Don Macfie, Ken Mackenzie, George Maier, David Mills, Douglas Park, Lloyd Smith, Dr David Stewart.

From Norway: Capt Erik Dokken, Capt Georg Evensen, Cato Guhnfeldt, Maj Gen Christian Kaldager, Capt Nikolai Slettevol.

From New Zealand, Australia and the Pacific: Therese Angelo (RNZAF Museum), Frank Chartres, Jim Dorman, Denys Jones, Capt Ron Gillies, Noel Hollé, Capt John Laing, Capt Phillip Mathiesen, Capt Lloyd Maundrell, Neville Mines, Capt Bryan Monkton, Chris Murray, David Murray, Kathryn Patterson (NZ National Archives), Norman Simpson, Brian Smith, the family of the late Harry Williams, Harry Woolnough, OAM.

Foreword

Captain Ken Emmott

I am honoured to be asked to write the foreword to Peter's book. Perhaps I am somewhat biased, having been the only remaining qualified pilot of this magnificent Sunderland flying boat for the past eleven years, but I have found the story of the life of this wartime aircraft completely fascinating and absorbing. I hope that others, whether involved in aviation or not, will also find it so. I am impressed by the depth of research and tremendous effort which Peter has put into the project, as can be judged by the huge list of sources quoted for the book.

The book is also, unwittingly, a comment on the initiation into aviation of Peter Smith himself. Originally he had little knowledge of, or connection with, the world of aviation, and in this book he takes the reader through his development into becoming the leading expert on *Islander* itself and to his acquiring license qualifications as a fully fledged ground engineer.

Peter is the type of person who pulls no punches! He is prepared to state his views face to face and in no uncertain terms to the highest in the land. This on occasions has led to trouble in the camp, but one can always rely on his having no ulterior motive of any kind. His aim has always been to put forward what he considers is the best course of action for the sake of his baby — *Islander*.

For many years now, he has been utterly devoted to the boat, living alongside it, sleeping aboard whenever it was on the water, working long hours on every aspect of the aircraft's maintenance including engine overhaul, the techniques of metal working, electrics, hydraulics, refuelling, launching and recovering — all of which over the years he has learned to master — and all this starting from an initial ground zero in terms of specialised knowledge.

For myself, having spent several years on Catalinas during the War, followed by the good fortune of having obtained employment with BOAC on Hythes, Plymouths and Solents immediately afterwards, until the flying boats were phased out in 1950, it was a thrilling experience in 1981 to be asked by Edward Hulton to take over the captaincy of *Islander*. I well remember that unique sensation upon launching, of the boat bobbing and nodding through the waves as she was taken out to the buoy. All flying boat men will remember this and it brought back for me all those memories of flying off water — there is nothing to match it on landplanes.

It was with some apprehension and tummy butterflies that, with Colonel Labonde (no experience on flying boats) of the French Air Force as co-pilot, we started up the four engines and taxied out into the Solent for engine and general shakedown trials. These were all successful and indeed it was on the first high speed run, with Engineer Bill Mares pounding me on the shoulder and pointing to the airspeed indicator that, despite my throttling back partly, the boat was thrown off by a wave and 'flew' at a height of about three feet, until I put her back in the water. Everyone on board was delighted by the experience — an involuntary manoeuvre on my part!

The transition from Boeing 747s to a Mk V Sunderland can only be described as a fantastic shock to the system. Since that time, with Mike Searle as co-pilot, Geoff Masterton as Flight Engineer and, need I say it, with Peter Smith as bowman, trouble-shooter and general factotum, we have had some interesting and exciting flying experiences together.

Captain Ken Emmott at the controls of ML814

(Photo Graham Playford)

We have made lots of friends, not least numerous Air Traffic Controllers in the south of England who have always wanted a closer look at the Sunderland as we passed by — so much so that on average an expected flight time of an hour would usually take two!

Contrary to a popular view of the CAA we have also made many friends there, all of whom have been most helpful in awarding the necessary certificates, dispensations etc., required to keep the boat flying. For me of course it has been a very fortunate and rewarding way to end my flying career.

Regrettably for all flying boat enthusiasts on this side of the Atlantic, the high cost of maintenance and hangarage along with other factors, have obliged Edward to offer the boat for sale. This process itself has been something of an epic — with offers and counter offers and much excitement regarding a future for the boat in exotic island paradises — all of which finally came to nought over a period of some three years.

Now at last *Islander* is to move to a new home in Florida, to be maintained and hopefully flown as part of the collection owned by the well-known American aviation enthusiast, Kermit Weeks. This will probably be the final great adventure for this wonderful, last remaining, flightworthy Sunderland. We will all be sad to see her go, but have the quiet satisfaction of knowing that she will continue to fly and will be looked after with all the care and attention that she deserves.

Chapter one

March 1944 and the Battle of the Atlantic

Sometime in mid-March 1944, the big doors at one end of the 300 ft. bay in Short and Harland Ltd's Belfast factory slid open, and out was towed the sixty-third Sunderland flying boat to be completed there. The Belfast factory was one of four producing the famous Short Sunderland flying boat during the Second World War.

On the hull of the new flying boat, just under the tail fin, was its identifying serial number — ML814. It would be nice to know a little more about the first days of ML814...

When exactly did it roll out of the factory? When was it launched? When was its first test flight? Which of Shorts' famous test pilots was at the controls?

However, despite Shorts being the world's oldest aircraft manufacturer — and still today they build aircraft in the same Queen's Island, Belfast, factory — they did not bother to keep such records. It is not possible to locate any of these details for ML814. Sad from the historical viewpoint but, under the pressures of wartime production, understandable.

Besides, who in the Shorts factory could possibly have dreamt of the career which lay ahead of ML814? Who could have known that she would go on to have a far longer, wider and more varied life than any other Sunderland, possibly than any other single aircraft of any type?

At the time ML814 was built, it was considered good going if a Sunderland lasted a few years, perhaps saw out the war. Yet not only was this flying boat destined to survive the war with three different squadrons, it was then to go on and play a role in the rehabilitation of war-torn Northern Europe.

Later it was to be a guardian of the tropical expanses of the South Pacific, where it brought the inhabitants of remote atolls within reach of medical attention for the first time.

Next, the military machine became a luxurious civil airliner, providing a vital link between a remote island community and the Australian mainland. Here the flying boat was no mere means of transport, but was essential to the community's existence. Even today, almost 20 years since the flying boat's departure from the island, people there still talk fondly of "their" flying boat.

And in 1993, after circling the globe to return to its native United Kingdom, Sunderland ML814 still flies, enabling a new generation to admire the graceful beauty of one of these legendary 'ships of the air' being put through its paces. If ever an aircraft was destined to play a major role in the lives of a great number of people, the world over, it was ML814.

But first we must return to the Belfast of March 1944. Newly completed Sunderlands normally spent a day or two on the hardstanding beside Belfast Lough. Here fuel flow checks would be carried out, initial engine runs made and any necessary adjustments effected. That completed, she was ready for wheeling down the wooden slipway and towing out to a mooring.

Sunderland Mk IIIs in the general assembly area at Short and Harlands, Queen's Island, Belfast, during mid 1944.

(Photo Short Brothers PLC, SU 400)

There, she would probably have joined a number of other Sunderlands, swinging at their buoys against a backdrop of the construction slips in sister company Harland and Wolff's Musgrave Yard. Amongst the vessels on the stocks there at the time were five new aircraft carriers for the Royal Navy.[1]

The group of white, four-engined flying boats looked deceptively peaceful at their moorings on the lough; more like their pre-war airliner relatives than deadly warplanes. Early in the war, Sunderlands had in fact been painted in a camouflage colour scheme, until a ministry scientist looked into the matter. To a lookout on the deck of a U-boat, he realised, the least apparent colour for a predatory aircraft was white. Man had finally discovered this fact, long since learnt by sea birds in the process of evolution. From then on all Sunderlands were white, albeit with grey upper surfaces to make them less obvious against the sea when viewed from above.

Once ML814 was on her mooring, the handling squad would have descended upon her in their work boat to remove the beaching gear. Lifting tackle would be fitted to an eye on the underside of the wing, fixing pins removed, and the upper end of one of the heavy, steel-box legs lowered, until it lay floating on the water. After the two main legs had been towed ashore, there was just the easier tail trolley to release and float from under the stern and the job was done.

Then would come a number of test flights, all quite short, perhaps ten or fifteen minutes each. Under wartime production conditions, no more time than necessary would be wasted here. The prime aim of these flights, apart from detecting any unexpected problem, was to check and adjust control surface rigging.

Ailerons tended to be tricky on Sunderlands, and the fixed trim tabs would be tapped up or down as necessary between flights. If more than a small adjustment (5°) was needed here, an aileron would have to be changed — this was a relatively common occurrence.

Another view of the general assembly area (300 ft bay).

(Photo Short Brothers PLC, SU 384)

As soon as the test pilot was happy with the Sunderland's handling, he would sign the form 1361 to certify that the aircraft was proved in the air in all respects. Next the company's Inspector-in-Charge would sign the Certificate of Final Inspection, and the new flying boat was ready for hand-over to the RAF. In the case of ML814, this occurred on 25 March 1944, the first date for which any record exists.

From Belfast, ML814 was flown across the North Channel of the Irish Sea to Wig Bay, near Stranraer, on Scotland's West Coast. This was the RAF's major storage and servicing base for flying boats at that time. Until the start of the war, Calshot, near Southampton, had been the hub of flying boat activities — training, maintenance and so on. Once the war started it was realised that the South Coast was vulnerable to air attack, and a number of new, dispersed bases were established.

Wig Bay, as it happened, was conveniently located just across the water from Belfast. On at least one occasion, when some Sunderlands were in a dubious state of airworthiness, they were even taxied across from Scotland to Ireland. ML814 arrived at the base, on the somewhat desolate looking western shore of Loch Ryan, on 26 March 1944.

Before she could go into squadron service, there was a considerable amount of final fitting out yet to be done by 57 MU, the RAF Maintenance Unit at Wig Bay. The boats left Shorts' assembly line fitted only with communications radio equipment. There was still further navigational radio equipment to be fitted by the RAF, not to mention the all-important radar, plus a great variety of military stores.

As with all Sunderlands delivered to the RAF during the war, it was a foregone conclusion that ML814 would be issued to one of the many squadrons of Coastal Command. There had been a few exceptions; a handful of boats had gone out to Australia and New Zealand, some had gone to BOAC for use as transports, and later in the war some went to the RAF for use in the Pacific theatre. But the vast majority of Sunderlands, for most of the war, went to Coastal Command.

Coastal Command was one of four commands making up the Royal Air Force, as set up during the re-structuring of 1936. It was an unusual force, almost a maritime air force of its own. It was a force of air-sea specialists, administered and staffed by the RAF, yet under the operational control of the Royal Navy.

The Command's personnel, at least in its early days, were typified by their Air Officer Commanding-in-Chief, Frederick Bowhill. After passing out from the training ship *Worcester*, he spent sixteen years in the merchant navy, transferring later to the Royal Navy, and later again to the Royal Naval Air Service. His two most prized personal documents were a Royal Aero Club Pilots Certificate (number 397), and the certificate of an Extra Master, Square Rigged.

Many of the older pilots and aircrew of Coastal Command at the start of the war were 'old salts', who had come into flying via the Royal Navy. They looked on themselves as an elite. Referred to as members of the 'Flying Boat Union', they encouraged a certain mystique around themselves, as demonstrated by their habit of never cleaning their cap badges or buttons. With continual salt water immersion in the old open cockpit flying boats, the experience of flying boat crews was supposedly determined by the degree of verdigris on their badges!

Of course, by the time ML814 was entering service in 1944, these old traditions had been diluted by the huge influx of new blood over the war years. Yet there was still very much a feeling amongst flying boat crews that they were part of a fraternity, whose special skills set them apart from the crews of mere landplanes.

To understand something of the role ML814 was likely to play in the Second World War, it is necessary to know a little of the battle into which she was to be thrust — the Battle of the Atlantic. This was the longest running and most vital battle of the whole conflict. The first shots were fired on the opening day of the war, and the battle continued until after peace was declared in the European theatre. To tell that story here is impossible, and there are many books available on the subject. However, to try to 'set the scene', I will endeavour to briefly tell of some of the highlights, particularly those relevant to Sunderland operations in Coastal Command...

The Battle of the Atlantic was the battle to protect the merchant vessels supplying Britain's needs. These supplies were required both to maintain the war effort and to keep the population alive — cut off these supply routes and Britain would fall. It was to become almost entirely a battle against those lethal raiders of commerce, the U-boats, although at the beginning of the war the Allies did not seem to realise this. The Royal Navy considered that the greatest threat came from German surface raiders.

For Britain, it seemed, the lessons of the First World War had not been learned. The U-boats had almost brought her to her knees in that war, and they were to do so again in the second war. The Royal Navy considered that the development of Asdic in the 1920s had all but removed the submarine menace. Utmost faith was being put in this technical advance, which detected submarines underwater by bouncing sound waves off them.

A small anti-submarine school had been set up in 1935 (HMS *Osprey*, at Portland), and its few graduates were only too aware of the problems facing them. They were far more aware of the limitations of their Asdic sets than were their superiors, and very often found that when they were aboard operational destroyers, their Commanders tended to consider the role of the anti-submarine specialist as almost irrelevant.[2]

New Mk III Sunderlands swinging at their buoys on Belfast Lough, against the background of Harland and Wolff's busy Musgrave Yard.

(Photo Short Brothers PLC, 80374)

No real technical or strategic developments were being made, and the Anti-Submarine Branch was undoubtedly the Cinderella of the specialists of the navy's Executive Branch. In short, the Navy was totally unprepared for what was to come.

From a German point of view, matters were rather different. In 1935, Germany had formally abrogated the Treaty of Versailles, which had prohibited submarines for the German Navy. Despite these restrictions, developments in submarine design had already been made by building German designed boats in foreign yards. After 1935 production went ahead apace in Germany.

In charge of German submarine activities was Karl Dönitz, a career naval officer who had commanded a U-boat in the First World War. Following a successful attack on a convoy, his U-boat was attacked by the Royal Navy and he was taken prisoner for the rest of the war. His experiences no doubt stiffened his resolve to 'get even' with the British when the Second World War came around.

Contemporary British propaganda painted him as something of a fool, as a land-lubbing admiral who hated the sea. He was, however, a very wily adversary who enjoyed the full respect of his U-boat commanders and their crews.[3]

Prior to the outbreak of war, Dönitz had estimated that with a fleet of 300 U-boats he would easily be able to bring about the total collapse of Britain. Perhaps luckily for Britain, however, Dönitz did not get everything his way. Hitler, not particularly interested in matters naval, starved him of the necessary funding. In September 1939, Dönitz had just 54 U-boats, and only half of these suitable for operation in the Atlantic.[4]

Nevertheless, he was quickly able to give the Royal Navy a bitter taste of what was to come. There was no period of Phoney War in the Battle of the Atlantic. On the very first day of war the SS *Athenia* was sunk by a U-boat.

Our vulnerability to German submarine attack was most devastatingly demonstrated in mid-October, with the loss of the battleship HMS *Royal Oak* in the supposedly secure anchorage at Scapa Flow. U-boat ace Günther Prien skilfully found his way into the harbour, pressed home his attack, then escaped, leaving the Royal Navy to mourn the loss of eight-hundred officers and men.

Coastal Command had, of course, been preparing for the inevitable over the months leading up to the outbreak of war. When war came, however, it found itself little better prepared than the navy.

The stringent military economies of the inter-war years had left the Command with a largely obsolescent force of aircraft, despite considerable expansion over the two years prior to the war.

Ten squadrons of Avro Ansons made up the backbone of the force, and they were not particularly suited to the role. Of the six squadrons of flying boats, four were old biplane boats, Saro Londons and Supermarine Stranraers. The few squadrons with the new Sunderland flying boat, which had come into service about a year before, were far from up to full strength.

The only other modern aircraft in the Command was the American-built Hudson. This was a derivative of the twin-engined Lockheed airliner. One squadron was working up, and orders had been placed for as many aircraft as Lockheed could supply. Thus equipped, Coastal Command took on its wartime role of the aerial protection of Britain's coastal waters and of the shipping lanes leading to them.

For the first half of 1940, Allied merchant shipping losses averaged 80,000 tons per month. Disconcerting, certainly, but not sufficient to be a real threat to Britain's existence. In these early stages, Coastal Command aircraft patrolled the North Sea, traversed by U-boats moving to and from the North German and Norwegian bases.

Longer range aircraft, of which the Sunderland at that time was the only one, patrolled out into the Atlantic with the convoys. The one lesson which had been remembered from the first war was the advantages of the convoy system, and this was introduced for merchant vessels right from the start.

A total change in the situation was brought about following the French armistice of June 1940. Dönitz immediately commenced the work of converting the French Atlantic coast ports into U-boat bases. Within weeks Lorient was ready for use by submarines, and the four other bases developed along the coast were in action soon after.

The opening of these bases changed the situation overnight. The North Sea became largely irrelevant, as U-boats could move from their new bases across the Bay of Biscay and directly into the Atlantic. For the second half of 1940, U-boat destruction of merchant shipping was to average 240,000 tons per month. Later, the German crews were to look back on this period as 'The Happy Time'.

Coastal Command aircraft seemed to be doing little but harass the U-boats, and up to this time they had only achieved several successful 'kills'. Air Chief Marshal Bowhill pressed for technical innovations to help redress the balance. By mid-1940, modified naval depth charges were being used from aircraft, replacing the largely ineffective anti-submarine bombs. That year also saw the first experiments with airborne radar to detect submarines.

Technical developments brought a series of pendulum swings in which alternate sides had the upper hand. As the British made their ASV (aircraft to surface vessel)

A Mk III Sunderland taking off from Belfast Lough on a test flight.
(Photo Short Brothers PLC, SU 315)

radar effective, so the Germans developed the French-manufactured Metox receiver, to detect the radar emissions.

The RAF developed the Leigh Light, which allowed the illumination and attacking of a surfaced submarine at night, after it had been detected by radar. These developments made life particularly difficult for U-boats during their crossing of the Bay of Biscay.

As a result, the U-boats switched their attack areas to mid-Atlantic. The Sunderland, despite its 14 hour endurance, could only effectively patrol up to 600 miles out into the Atlantic. When the American Catalina flying boat was brought into use by the RAF, it allowed a reach of 800 miles; but there was still the 'mid-Atlantic gap', where the convoys were unprotected by aircraft and at the mercy of the U-boat. The answer proved to be longer range aircraft, the American Liberator bomber for Coastal Command, and small escort aircraft carriers sailing with the convoys. Both, however, took time in coming.

The Royal Navy was also getting its own priorities right, with more and better escort vessels, and new weapons such as the hedgehog anti-submarine mortar. Perhaps their greatest successes were due to the efforts of Captain Gilbert Roberts.

A retired officer called back to investigate anti-submarine strategy, Roberts set up the Liverpool anti U-boat school and was responsible for radically improving the navy's success rate against this enemy.[5]

The entry of the United States into the war had its effects as well. At first, the US Navy was not as well-equipped for the anti U-boat role as the Royal Navy had been at their entry into the war. Dönitz switched many of his boats to the easy pickings of the Western Atlantic. But, by mid to late 1943, the US Navy was making its presence felt, particularly with its escort aircraft carriers.

In early 1943, Allied losses reached their worst levels ever, and things began to look particularly grim. May, however, was to be the turning point of the Battle of the Atlantic. Technical developments again played an important role. HF/DF (Huffduff) radio equipment on ships, and a new centimetric radar on aircraft, both enabled the detection of submarines, without them even realising that they were being hunted.

During one two-week period in mid-May ten convoys, including some 370 merchant ships, passed through the mid-Atlantic danger zone for the loss of only six ships. In return Dönitz had lost thirteen U-boats — seven directly to aircraft, four to naval escorts, and two to aircraft in conjunction with naval forces.

Dönitz was at a loss to know how the Allies were locating his U-boats, but he knew that he was losing the battle. On 24 May he ordered his commanders to withdraw from the mid-Atlantic, some to transit to the Azores area, others to return to France.

So by April 1944, when the new Sunderland ML814 completed her fitting-out with 57 MU at Wig Bay, and was ready to join the fray, some may have said that the Battle of the Atlantic was already over. But not many at Coastal Command Headquarters believed this, and Grossadmiral Dönitz most certainly did not. He could see technical developments on their way which would swing the pendulum back in his favour, and allow him even yet to achieve his grand aim.

Footnotes Chapter 1.

[1] *Shipbuilders to the World* — M Moss and J R Hume
[2] *The U-boat Peril* — Bob Whinney (p 42)
[3] *Coastal Command Leads the Invasion* — Wilson and Robinson (p 16)
[4] *The Underwater War 1939-1945* — Richard Compton-Hall (p 16)
[5] *Captain Gilbert Roberts, RN* — Mark Williams

Chapter two

Into Service with 201 Squadron

ML814 was duly allocated to an operational squadron, Number 201, based at Pembroke Dock in Wales. No. 201 was a unit very much steeped in the traditions of Coastal Command — in fact it was older than the Royal Air Force itself! It had been formed in October 1914 as No. 1 Squadron of the Royal Naval Air Service, the designation changing to No. 201 on the formation of the Royal Air Force in 1918. The Squadron had been employed on coastal patrols, and one of their aeroplanes is believed to have been the first ever to attack a submarine.

Between the wars, the Squadron was based at Calshot and equipped with the big biplane flying boats of the period. In fact, when it was moved to its war station at Sullom Voe in the Shetlands during August 1939, it was still using old Saro London boats. These were replaced by Sunderlands during 1940, and the Squadron was to move firstly to Northern Ireland before coming to Pembroke Dock.

On 24 April 1944, a skeleton crew of four under Flight Lieutenant Aubrey Poole were ferried out to ML814 at its mooring on Wig Bay, where they checked over their Squadron's newest addition prior to the flight back to Pembroke Dock. All seemed well, and at 1505 the boat left her mooring and taxied out for take-off.

The flight was a short one of just under an hour-and-a-half, over water all the way and passing just off Holyhead on Anglesey, and further south skirting Bardsey Island. As the crew picked up the coast of West Wales at Strumble Head, then swung in over Milford Haven, there was a feeling of "it's good to be home".

After the barren countryside around Loch Ryan, Milford Haven looked very homely. To either side was farmland, with narrow lanes linking the towns and villages which had become familiar landmarks to the crew. The Haven itself was a long narrow inlet of deep water, protected on all sides and thus a magnificent harbour. The largest of vessels could enter at any state of tide, and during the war it was always full of naval vessels, merchantmen and flying boats. Since the war it has become the site of several large oil refineries, taking advantage of the deep harbour to berth enormous supertankers.

But in 1944, with three squadrons in residence, RAF Station Pembroke Dock was reckoned to be the largest flying boat base in the world. Half way up the Haven, the incoming Sunderland passed Angle Bay on the southern shore, with its night take-off area and rows of flying boat moorings or 'trots'. Even though quite a number of the boats were either flying on exercises or 'up the slip' for maintenance, ML814 skimmed past between twenty and thirty other white Sunderlands swinging at their moorings.

Ten miles up from the sea they arrived at Pembroke Dock itself, jutting out from the southern shore. Here, just off the station, Aubrey Poole put the new boat gently down in the alighting area and taxied it to his allotted mooring. Once the Sunderland was moored and its locks and covers fitted, a dinghy drew alongside and took the crew across to the jetty. Along the waterfront from the jetty were a series of small dock basins, ideal for the myriad dinghies, pinnaces, bomb scows, and other small craft needed by a large flying boat base. On the hard standings between the docks stood a number of Sunderlands, looking somewhat clumsy on their beaching gear.

Ashore, RAF Station Pembroke Dock had an atmosphere of its own, quite different from that of most other operational RAF stations. It was the architecture, the sense of maturity. The majority of the buildings did not date from the aeroplane age, but were much older. The station had only been transferred to the RAF in 1930, its previous history as a Royal Dockyard being evident from the solid stone naval buildings, mostly Georgian.

The RAF had started to make its mark as well, naturally. A wide swathe of concrete cut up from the slipway to a large open field. At the top end of this concrete apron stood two huge hangars, one to either side. Beyond the open field, variously used for sports and for flying boat parking, was the most grandiose of the new air force buildings — the red brick officers' mess.

Although ML814 had now been delivered to its first operational squadron, it was still not ready for service. The following day the same skeleton crew who had collected the new boat from Wig Bay were scheduled to deliver it to Calshot. There were a number of 'mods' to be carried out before she was ready for battle. And although 201 Squadron had left Calshot at the start of the war, their flying boats still returned to the old base on Southampton Water for heavy maintenance.

For Aubrey Poole's crew, the flight was welcome news. Pembroke Dock may have been an excellent base, but to many of the personnel stationed there, the town of Pembroke Dock was looked upon as the end of the earth. Dockyard towns were not normally known for their magnificence, and this one was small, had become run-down and, as a final blow, was blitzed.

So it was a cheerful group who rode out in the dinghy around lunch time for the flight to Calshot. With a bit of luck they would be able to arrange a 'night on the town' in Southampton, before returning in a different aircraft the following day. And not being a full crew, perhaps their routine was somewhat disturbed as they rushed through the pre-flight checks.

Soon all seemed in order, the two outer engines were started, and the mooring slipped. After the normal warming up of the engines and the power checks as they taxied, all was ready for take-off. The boat swung around into wind, and the throttles were pushed forward. With only a light fuel load and little else on board, the Sunderland was quickly up on the step and planing. The pilot glanced down at the instrument panel — there was no reading on his air speed indicator, none on the second pilot's either. Someone had forgotten to remove the pitot head cover.

Of course they could have stopped the run, removed the cover, then started all over again. But it would have wasted time, and far worse, would have caused a red face or two. So the flight continued, and Calshot was successfully reached without that most vital instrument, an ASI.

From the air, Calshot was an obvious landmark. The RAF station stood on Calshot Spit, a finger of shingle extending some three-quarters of a mile into Southampton Water from the western shore, at its junction with the Solent. Situated some six miles down Southampton Water from the docks, it was a familiar sight to anyone leaving the port by sea.

With such a strategic location in guarding the port of Southampton, Calshot Spit had for centuries been associated with the defence of the realm. In the reign of Henry VIII a circular castle had been built at the tip of the Spit, and it still dominated the site. Around the castle the Royal Navy had constructed one of its earliest Air Stations, later to become RAF Station Calshot. The station now amounted to eight or so large hangars.

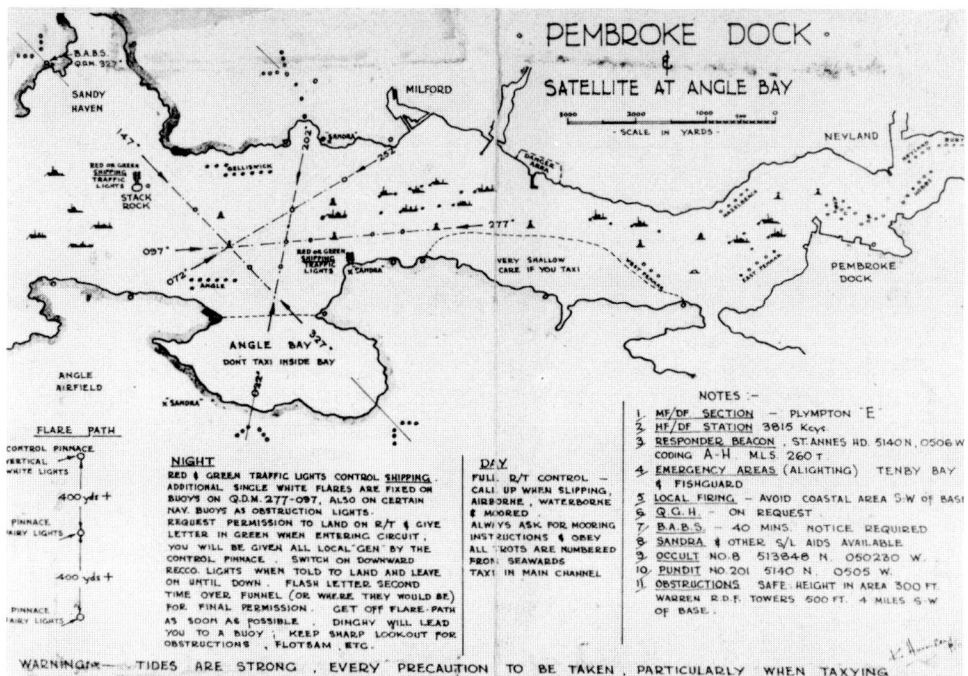

A 1944 landing chart of Pembroke Dock, also showing Milford Haven down to Angle Bay.

(Chart via George Maier, ex 422 Sqn.)

As Aubrey Poole and his crew approached Calshot on that day in April 1944, there were signs of the impending invasion of Europe everywhere. On the Solent foreshore, just along from Calshot at Lepe, were some of the huge concrete caissons known as Phoenixes. They were to form a small part of the floating harbours to be towed across to the invasion beaches.

Up Southampton Water itself toward the docks there were great numbers of other weird, unrecognisable floating structures, not to mention hundreds of landing craft and merchant vessels of all types. In places, they completely covered the surface of the large dock basins. Floating above all of this were many barrage balloons, so the Sunderland had to pick its way carefully through them to approach the alighting area.

Once ML814 was safely down and moored, and the crew taken ashore, the beaching party started work. With so much naval activity around, Southampton Water was not a good place to have a big Sunderland riding at its mooring for long. So the new aircraft was brought in to the beaching buoy, just off the North Slipway. There the two large beaching legs were towed out to her in turn.

Beaching was a tricky job, especially if there was anything of a sea running at the time. Working with the big heavy legs from open boats was not easy, and the wash from a passing vessel could create havoc. But to the beaching party it was all a routine matter, and carried out without a second thought. Finally, the wooden tail trolley was floated out and fixed under the Sunderland's stern, then she was hauled up the slipway. Little did anyone think at the time that this same Sunderland would regularly be coming up and down the Calshot slipway in 48 years' time!

Unfortunately, there is no documentation to show exactly what modifications were carried out on ML814 at Calshot that April, but it would seem to have been some work to do with the armaments.

By now ML814 was painted up with her squadron identification letter R, and squadron codes for 201, NS. Thus the large letters NS - R appeared on either side of her rear hull, separated by the RAF roundel. She would normally be referred to as R/201, or verbally, in the phonetic alphabet of the period, as R for Robert. Squadron personnel, incidentally, had not the slightest doubt that the code NS had been allocated to No. 201 to stand for *Nulli Secondus*.

A crew of eleven was allocated to R/201, including their captain, Flying Officer Dave Easton. Flying boat crews appreciated the fact that they had their 'own' boat, as opposed to landplane crews, who were merely allocated any available aircraft for a particular sortie. It encouraged a pride in caring for the boat and keeping everything ship-shape, which often extended to crews fitting out the wardroom with curtains and hanging photos on the walls. Others delighted in 'souveniring' crockery and similar items from local cafes, so that many a Sunderland galley was equipped with much more than its RAF issue crockery.

ML814's first permanent captain, Fg Off D A Easton, had worked in a bank (Morgan Grenfell) before the war. In 1938 he had joined the RAF Volunteer Reserve, flying Tiger Moths at weekends from a grass field at Gatwick. He was called up in September 1939. By 1941 he was a Blenheim pilot and with his squadron, No. 248, took part in the Battle of Britain. For Dave Easton and his crew, their first outing in R/201 was an easy and straightforward one, a test flight from the base at Mount Batten, Plymouth on May 9th. Apparently the boat had been transferred there from Calshot.

Mount Batten, like Calshot, was another long established flying boat base. Home to the Sunderlands of No. 10 Squadron, Royal Australian Air Force, it suffered from having to share Plymouth Harbour with the navy. However, in the case of bad weather at Pembroke Dock, it would often be required as a diversion, so 201 Squadron crews needed to know it well. The following day they carried out a daylight base familiarisation flight, taking in the Scillies and Calshot. The Scilly Isles weren't a base in the real sense, just an emergency alighting area. Located as they were at the toe of England, they were first landfall for flying boats returning from patrol in the Bay of Biscay.

Thus, for the odd boat suffering from severe battle damage, perhaps low on fuel or with one or more engines out, these small dots on the ocean were a very welcome haven. This familiarisation flight brought ML814 and her new crew back home to their base at Pembroke Dock. Dave Easton was later to comment on what a pleasure it had been to him and his crew to be allocated ML814: "brand new, all tight and fresh like a new car."

Two days later R/201 was off on her first real exercise, night flare dropping and radar training, over Carmarthen Bay. With 201 Squadron — in fact with all the squadrons of Coastal Command's 19 Group, which covered the southern UK and its surrounding sea areas — training was what life was all about in May 1944. Everyone knew that the invasion of Europe wasn't far off, it was just that no one actually knew when or where it would happen.

It was in preparation for this great event that 201 Squadron had been moved from its previous base to Pembroke Dock. Castle Archdale in Northern Ireland had given them better access to their patrol areas, far out over the Atlantic. But when the invasion came, no one was in any doubt that the U-boats would be out in force to create huge losses amongst the vulnerable invasion fleet. Thus 201 came to Wales in March, to put them near the Bay of Biscay and the Channel approaches.

The pilots' positions in a Sunderland. Captain to the left, second pilot to the right.

(Photo Short Brothers PLC, SU 672A, via Ron Parsons)

Along with the move, 201 was re-equipped with the new Mk IIIA Sunderlands — ML814 having actually been one of the last to arrive. The Squadron now had twelve of these new boats operational, plus two in reserve. They differed from the older Mk IIIs only in the more sophisticated electronics which they carried, in particular the new ASV Mark 3 radar.

This radar update was apparent at first glance to the knowing observer, because gone were the great arrays of stickleback aerials of the old radars, replaced by two neat underwing bulges, each housing a scanner. The equivalent installation on a landplane only needed one scanner, but you could hardly put a scanner under the hull of a flying boat.

The greatest advantage of the new radar was not its much superior performance, but the fact that the U-boats' Metox receivers would not be able to detect the approaching attacker. The Germans were certainly aware of its existence, and were working towards the perfection of a detector, but were not yet able to detect transmissions on its ten centimetre wavelength.

It had always been a bone of contention between Coastal Command and Bomber Command of the RAF, that Bomber Command was usually given the new technical developments for their use first. From Coastal's point of view, a piece of new electronic gadgetry was merely a temporary advantage against the U-boats — it would only be a matter of months before the U-boats became aware of its existence, and German scientists produced a device to nullify its effectiveness.

Bomber Command aircraft were far more likely to fall into enemy hands than were those of Coastal, and the Germans had already retrieved one of these new ten centimetre radars from a crashed Stirling bomber in Holland. But in the meantime, while they still had the advantage, Sunderland crews were out training with their new ASVs.

As well as training on the radar, Dave Easton's crew practised with new high-intensity flares over Carmarthen Bay. It was mentioned in the summary of earlier stages in the Battle of the Atlantic, how the equipping of Wellington bombers with a powerful searchlight — the Leigh Light — was a major factor in the victories against the U-boats in 1943.

Until this time most anti U-boat patrolling by Sunderlands had been by day, but from now on it was to be largely by night. Sunderlands, however, were never fitted with Leigh Lights. New high intensity flares had been developed, and each boat carried up to 80 of the 1·7-inch flares.

To the old hands on the crews, these new developments were not really welcomed. There was a nostalgic yearning for the days when the captain had the feeling that everything was in his own hands; deck-level attacks on the U-boats, dodging the flak during the run-in, while the nose gunner did his best to take out the enemy's gun crew. Then the captain released his stick of eight depth charges by instinct, hoping for the perfect straddle — four depth charges to either side of the hull and a sure kill.

Officially, this was not to be any more, at least not for night attacks. Now the enemy was detected by radar. The radar operator directed the captain to his target, and the depth charges were to be released from 300 feet, under the control of the navigator operating a gyroscopically controlled bombsight from a position under the nose turret.

Target illumination was provided by manually pumping the new electrically-fused flares out of a chute in the stern of the Sunderland at one second intervals. In reality, however, many skippers and crews continued to bomb visually in preference to using the new sight.

Apart from co-ordinating all these new skills, just getting used to the flares was a problem for the captain. Despite being dropped from the stern, the intense brilliance of these flares was such that they seemed to leap out from ahead of the flying boat. At first pilots instinctively flinched, this being transmitted to the controls and moving the boat from its correct course at the vital time. It was just a case of practice, and that is what ML814's crew was about on the night of 12 May.

On 14 May it was the real thing for the first time, Sunderland ML814 was off on an operational patrol. There was no such feeling amongst the crew, of course. To them it was just another routine patrol, although it was nice to be doing it in their new boat. The likelihood was that it would be an extremely monotonous patrol, hour upon hour in their solitary Sunderland, without escort and without sighting the enemy.

This was the situation for all Coastal Command crews, whatever aircraft type they flew. Long hours of almost unendurable boredom, during which they had to try to be fully alert; interspersed — ever so infrequently — by brief spells of extreme danger.

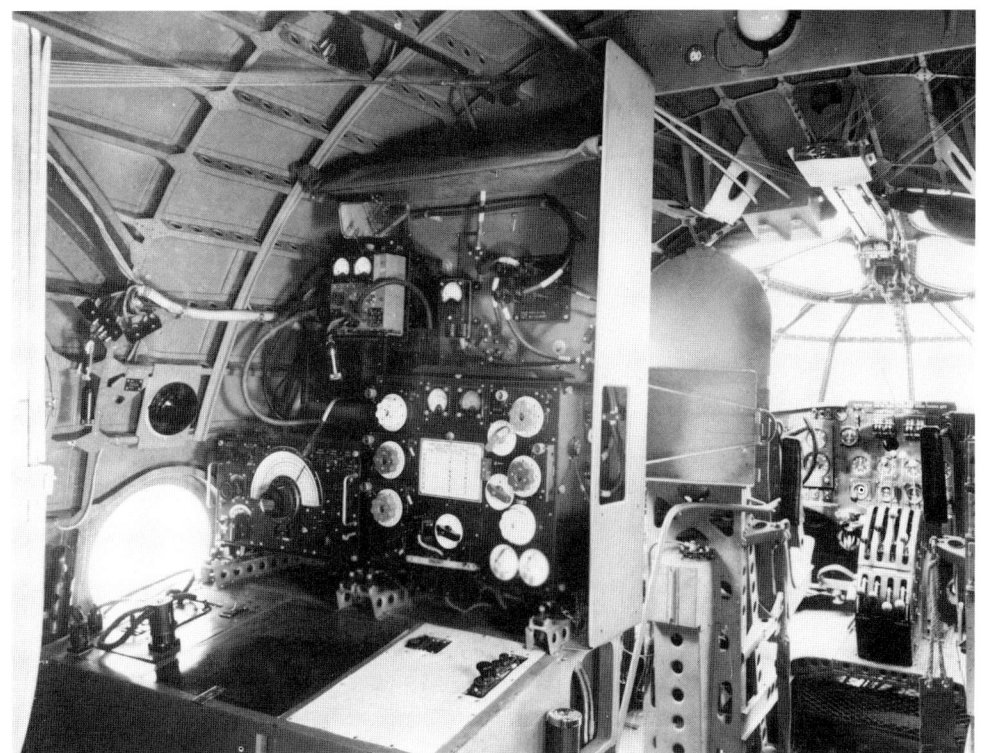

View forward on the bridge. Wireless operator's position to the left, with the pilots ahead of him.

(Photo Short Brothers PLC, 83611-4, via Ron Parsons)

Because of this, and because bringing a flying boat pilot to the necessary degree of skill and experience took far longer than on landplanes, Coastal Command made much heavier demands on their crews than did the rest of the air force.

Bomber Command crews rarely exceeded several months of operational service, and they could be counted lucky if they managed to survive this long. But a Coastal Command crew was required to fly on and on, until either they reached the set total of 800 operational flying hours, or till the day they headed out across the grey Atlantic in their big Sunderland, never to return. The remarkable thing was how many of the crews did manage to complete an operational tour, needing to make perhaps seventy or more sorties to achieve it.

True to the crew's expectations, the first operational flight of 14 May was very routine. It wasn't even a full night patrol, although a part of it was flown at night. At 1306 the Sunderland slipped her mooring and taxied out. Her take-off run was much longer than for the previous flights, with nearly twenty tons of flying boat, seven tons of bombs and fuel, plus the eleven crew. She struggled, nose high, to climb on to the step and clear those great waves of spray. Once on the step it seemed like miles as she thundered on, barely in contact with the surface as she built up to flying speed, then finally lifted ever-so-slowly away.

It was half past two the following morning before she was back, touching down on the flare path at Angle Bay. Thirteen-and-a-half hours, a usual sort of patrol length. In fact everything had been very usual.

The entry in the Squadron Operations Record Book read as follows: "Anti-submarine patrol. 14 May. ML. 814. R/201. Up 1306 Down 0227 (15/5) Nothing whatever was seen. On patrol 1626 off 2146. Area 48°N 10°W. 5/10 cloud at 1000 / 2000 ft. Average height flown 800 ft."

This patrol, down over the old familiar territory of the Bay, had taken Dave Easton's crew to the borderline of what was considered a danger zone. In the old days, when Sunderlands ranged over the wide expanses of the Atlantic, they met little except the weather to threaten them. But as they came to worry the U-boats more and more in the Bay, the U-boat men looked to the Luftwaffe for protection.

Protection was given in the form of patrols by the efficient, long-range, twin-engined Junkers 88 fighter. Armed with cannon, it should have been able to run circles around the slow, lumbering Sunderland in a fight. To make things worse they usually operated in pairs, or large packs. Certainly the Sunderland had plenty of guns, but they were all ·303s. The enemy had to be very close before these had any hitting power. Anywhere to the east of 10° west was considered likely territory to put your Sunderland in the sights of a Ju 88.

Occasionally the RAF sent out Beaufighters to take on the Ju 88s, and these met with some success. Many a Sunderland was lost to the Ju 88s, but some successes had been scored as well. Numerous Sunderlands had fought off one or two attackers, and several had dealt with even more and returned to tell the tale. The German fighter pilots developed a healthy respect for the big flying boat, nicknaming it the 'flying porcupine'.

Five days after their first patrol in R/201, Dave Easton's crew were bound for the Bay again on another anti-submarine sweep. This time it was a night departure, with all the difficulties that involved. The boat was taxied down Milford Haven the previous day, and left on the trots at Angle Bay, close to the night take-off area.

The crew would gather on the dark jetty, some laden with the rations just drawn for the sortie, eggs and chops, bread and beans. Now to board the waiting pinnace. They struggled down the steps, rigged up in cumbersome flying gear and boots, binoculars around their necks, and laden with charts and other paraphernalia just collected at their briefing. Almost inevitably it seemed to be windy, but thank God it was at least now May, and not too cold.

The pinnace made good speed down the harbour, and there was shelter from the wind, but it was still a thirty-minute trip. Some of the crew stared almost mesmerised by the dancing patterns of phosphorescence in the craft's wake. Although they had seen this sparkling light on the dark water many times before, it always held a strange fascination.

There was a bump as the pinnace came alongside the control tender at one end of the Angle Bay flare path, and the crew clambered up onto the deck, back into the wind. As the skipper checked the night's arrangements with the flare path officer, the others slid down into the open dinghy which had come alongside, heaping their equipment around them. There was just enough room, so a certain amount of precarious re-arranging of bodies was always needed, until everyone was aboard.

The coxswain opened the throttle and the bow lifted high, speeding through the darkness towards the waiting Sunderland. As he throttled back to bring his dinghy alongside, a light went on aboard the Sunderland, and the port entrance door opened. A duty crewman took a line made it fast to the cleat by the door.

Everyone piled aboard and went through into the wardroom and galley, where some of the equipment was dumped. Then they made off to remote corners of the flying boat, each busy on his own duties. There were bilges, guns, and ammunition to check. Covers and locks to remove. There was a sudden noise like a whining motorbike from outside — the APU in the leading edge of the starboard mainplane had been started to provide electric power.

View aft on the bridge. Navigator's chart table to the left, and beyond the front spar (the large diagonal tubes form part of it) is the rearward facing flight engineer's position. Through the "oven door" is access to the rear of the flying boat.

(Photo Short Brothers PLC, 87361-3, via Ron Parsons)

There were charts to set up, portholes to black out, and exactors to bleed. Down in the bows, in front of the wound-back turret, the moorings were released and the boat put on short slip. Wirelesses and intercoms were checked, signal colours arranged. The activity subsided, Dave Easton took his seat, and the rest of the crew grabbed their Mae Wests and moved up onto the bridge. The two outer engines burst into life in quick succession, the crewman in the bow released the short slip line, and the big boat moved off slowly across the dark water. Ahead were the three tenders, lit overall, swinging at their moorings to mark the take-off run. Once airborne, course was set for the patrol area — 45°N 10°W. Back into Ju 88 territory again, but at least they had not been making their presence felt in the area of late.

As the Sunderland droned slowly southwards towards the patrol area, and when those on watch in the turrets could just pick out those first hints of daylight on the horizon — the false dawn — the radar operator broke the silence on the intercom. "S.E. to captain, I've got a contact ahead, just off the starboard bow". Next came the report: "Range now three miles, almost dead ahead, and coming towards us. Pretty sure it's an aircraft." The Fire Controller climbed into position in the astrodome from where he had a 360° view around the aircraft. It was his job to co-ordinate the gunners' action and advise the captain if evasive manoeuvres were needed.

The first visual sighting came from one of the turrets. It was Sgt 'Nick' Nicholson. Nick was a typical dour Yorkshireman, reckoned by the rest of the crew to have the eyes of a hawk. "I can see him. I can see him." All eyes peered into the semi-darkness where they could just make out a twin-engined aircraft. Someone said "Beaufighter!". A laconic voice on the intercom commented: "Never seen bloody Beaufighter wi' black crosses on it before."

All turrets swung to train on the now identified Ju 88, everyone itching to open fire. Dave Easton told the gunners to hold their fire for the time being, to await any action from the 88. There was an uneasy silence as the crew contemplated the potency of his 20mm cannon armament against their own puny ·303s.

The Ju 88, a Sunderland's airborne enemy over the Bay of Biscay.

(Photo MAP)

To their surprise, the German aircraft circled the flying boat, waggled its wings as if in a gesture of 'Good Morning', and continued its flight northwards. Not a shot had been fired.

A bewildered and relieved Sunderland crew watched him disappear, then followed a discussion on the intercom, trying to explain the Ju 88's unexpected behaviour. The consensus seemed to be that perhaps he was one of their meteorological aircraft on the way out to do his work off Ireland, and quite likely not armed at all. Whatever the explanation, the crew soon settled down into the monotony of their patrol. Apart from the Ju 88, the patrol was another "nothing whatsoever seen".

This patrol was to be the only other one of the month for ML814. U-boat activity was light in the Bay. Their crews were still licking their wounds from the defeats of last May, and awaiting the arrival of the promised new craft which Dönitz believed would put him on top again. They too, like the Allies, were planning for the expected invasion. Over the past six months Allied shipping losses to the U-boats had averaged a mere 80,000 tons per month.

As part of his temporary measures, Dönitz was improvising. U-boats came in many shapes and sizes, although backbone of the fleet throughout the war was the 740-ton Type VIIC, the perfect attack boat. But all were diesel electric, which meant that they were not truly a 'submarine', more a submersible boat. Below the surface they depended on batteries — limited speed, limited endurance. It was in coming to the surface to charge their batteries, whether by day or night, that they were vulnerable to air attack.

Thus the fleet experimented with, and by mid-1944 were fitting, a new device — the schnorkel. This long tube, carried on the deck casing and erected when the boat surfaced, allowed it to proceed at periscope depth on its diesels, which breathed through the tube. There were difficulties, and the device was not entirely popular at first with the crews.

It necessitated very accurate depth keeping. If a wave came over the top of the schnorkel, or if it went below the surface, a ball valve slammed shut. The diesel was still running, gulping in air, so it took the air from within the boat. Pressure dropped rapidly, ears popped, eyes bulged.

But techniques improved, more U-boats were fitted with schnorkels, and to their crews the new device became a welcome alternative to a depth-charging by Coastal Command.

Only too aware that the invasion was not far off, but not knowing when or where it would strike, the Germans made their plans. Admiral Theodore Krancke, commander Naval Forces West, was responsible for the planning. Initially, all U-boats would remain in their reinforced pens in the Biscay ports, safe from bombing in any pre-invasion attacks.

Once the invasion itself was under way they would go out en masse. No matter what losses they suffered, Krancke believed that quite enough would get through the Allied defences to do enormous destruction amongst the vulnerable invasion fleets. Allied planners had an advantage, they knew that the invasion would take place on the Normandy beaches. There were two ends to the Channel through which the U-boats could threaten the invasion. But the eastern end was shallow, and would be blocked with mines. Could aircraft totally barricade the western end?

This problem was the responsibility of 19 Group, now under the command of Air Vice Marshal Brian Baker. Great consideration was given to the matter at Coastal Command headquarters, and a staff officer there, Flight Lieutenant James Perry, came up with a simple but promising plan.

On a map of the area to be blockaded, Perry marked out twelve rectangles of varying sizes. Each rectangle was drawn so that one of the various types of aircraft available to Coastal Command could fly around its circumference in an hour, and its radar surveillance sweep would completely cover the rectangle. Thus the problem of co-ordinating many different types of aircraft, with differing cruise speeds and radar coverage, was greatly simplified.

Trials of the system had been carried out to the south of Ireland in April, using the British submarine *Viking* and a 201 Squadron Sunderland. Those responsible had confidence that the system would work. Royal Naval escort craft would provide close-in protection for the invasion fleet, but it was hoped that the aircraft blockade would keep the majority of the U-boats out.

Because it was hoped that these rectangular patrols would 'bottle-up' the western entrances to the Channel, they became known as 'Cork' patrols. In theory, at least, the entire area between Ushant and Land's End would be sealed off by this endless chain of aircraft.

The crews, of course, new nothing of all these plans. But like their enemy, they were only too aware that the big day was rapidly approaching. It was obvious by all the training — training to the point where it wore them down. R/201's crew took her up on three more training flights in late May, two of them at night and one involving 'attacks' on a friendly submarine. The emphasis was on night training, as 201, along with two squadrons of Halifaxes, had been selected to concentrate on night operations with the new high intensity flares.

By the start of June, the intricately planned wheel of events which would lead to Operation Overlord — the Normandy invasions — was already turning. Ships laden with cargoes of men and equipment were leaving the more distant ports for the assembly area to the south of the Isle of Wight. Then bad weather caused a 24-hour postponement, and doubts beyond that. On the evening of 4 June, advised by RAF meteorologists of a likely break in the unseasonable storms, General Eisenhower made the dramatic and lonely decision to proceed — the invasion was to commence on the sixth.

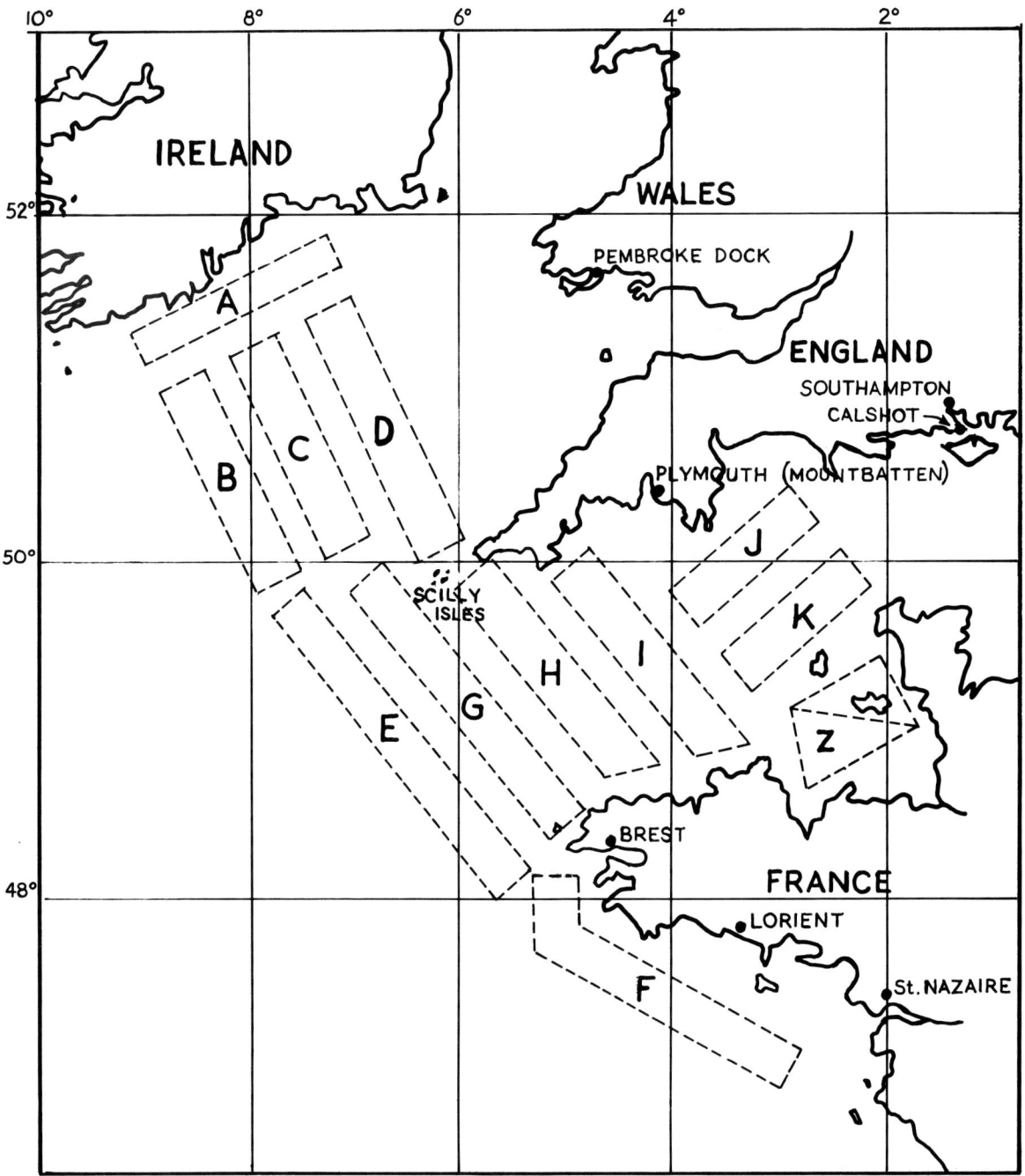

The sides of the boxes are the tracks flown by the patrolling aircraft.
This resulted in total radar coverage of the areas within and between the boxes.

Left: The Cork Patrol areas — Flight Lieutenant James Perry's simple plan for an air blockade of the western approaches to the Channel.

(Drawing Robin Allen)

All was set for the greatest invasion in history. But alongside it, not attracting the same newspaper headlines, would take place the greatest aircraft\submarine battle of all time. And in the midst of this battle would be ML814.

At 0513 hours on 6 June, 1945, German Naval Forces West flashed out an order to all U-boats of its Landwirt (anti-invasion) Group to come to immediate readiness. The boats already had their orders from Dönitz: "Every (Allied) vessel taking part in the landing, even if it has but a handful of men or a solitary tank aboard, is a target of utmost importance which must be attacked regardless of risk. Every effort will be made to close the enemy invasion fleet regardless of danger from shallow water, possible minefields, or anything else. Every man and weapon destroyed before reaching the beaches destroys the enemy's chances of ultimate success. Every boat that inflicts losses on the enemy while he is landing has fulfilled its primary purpose even though it perishes in so doing."

The first U-boats to set out against the invasion fleet left their concrete bunkers at Brest on the evening of 6 June. Fifteen boats set out, one low silhouette following another, until they disappeared out to sea. An impressive sight, or an eerily spine-chilling one, according to the point of view of the observer. Over the next day or two a total of 49 U-boats were to sail from the Biscay ports, about one third of them fitted with the new schnorkels.

Out over the Channel approaches, 19 Group was already waiting for them, patrolling the boxes in their Liberators and Wellingtons, in their Halifaxes and Sunderlands. That first night went well for 19 Group, their aircrews sighted U-boats on 22 occasions, and pressed home seven attacks. Five U-boats limped back into the French ports, damaged or badly shaken. The two kills were both down to Sunderlands. Despite the nearness of many patrols to the French coast, not a single Luftwaffe aircraft had threatened an Allied patrol.

Spirits were high amongst Dave Easton's crew as they slipped R/201's moorings at 9 o'clock the following evening, and taxied out for their first patrol since the invasion. There was no doubt that success did wonders for morale. Each crew member was thinking to himself "Will it be our turn tonight?" One of the two U-boat kills of the previous night had been down to S/201, Les Baveystock's crew.

Sadly, no such luck was to come to R/201. Their patrol was not a true Cork patrol in one of the lettered boxes. It was further south into the Bay. In fact the Sunderlands were not often used for the box patrols, but generally operated out to the south or west. It was mid-morning on the 9th when R/201 touched down on Milford Haven, completing another "sortie without incident".

It was almost beginning to look as though ML814 was one of the unlucky boats. Total score for 19 Group on 8 June was 142 sorties flown, 80 U-boat sightings, out of which six were attacked. For once the odds were actually in favour of a sighting, yet still that achievement evaded R/201's crew.

Two days later they were up again, this time on a daylight patrol. It was their first Cork patrol, and they carried out six circuits of their box. Weather was poor, and they were ordered to land at Castle Archdale as conditions had closed in totally at base. This order was later cancelled, and return was to Milford Haven as usual.

The mood in the mess at Pembroke Dock was quieter that night, however. Sunderland U, of 228 Squadron, had failed to return. At last report she had sighted a U-boat and was going in to the attack. Presumably she had fallen victim to the guns of her intended victim. It was the second boat 228 had lost in this way in two days.

A flying boat station, even in wartime, enjoyed a certain air of permanence due to the long time-period necessary to bring crews up to a good level of skill. The transitory feeling encouraged in Bomber Command, that crews were here today and gone tomorrow (and such a philosophy was needed if the crews were to retain their sanity), was not to be found at Pembroke Dock. Thus the losses, when they did come, were keenly felt.

The bad news was to strike even closer to home for 201 personnel. The following morning, the 12th, news quickly spread that S/201 had not returned. This was Baveystock's boat, successful in combat only four nights earlier, but less its usual captain on this occasion. Les Baveystock was on leave following the death of his father, so the 201 Flight Commander, Sqn Ldr Ruth, DFC and Bar, had taken his Sunderland and crew out. The Squadron Diary noted "...almost certainly shot down by the U-boat it was attacking. A most grievous loss."

Coastal Command may have been counting its losses, but there were many more losses to count across the Channel in the U-boat bases. By now a total of six U-boats had been lost to patrolling aircraft, and six more had limped back into port badly damaged.

On 12 June Admiral Krancke, admitting defeat, noted in his diary: "All submarines operating without the schnorkel in the Bay of Biscay have been ordered to return to their bases, as the enemy air attacks are causing too many losses and too much damage. Only if an enemy landing seems imminent on the Biscay coast are the boats to operate. They will remain under shelter in a state of readiness. . ." No 19 Group had reason to feel satisfied. Operation Cork was working.

When ML814 set out for her next Cork patrol, on 15 June, there was a new crew member aboard. This was Ron Harris, who had been posted to 201 Squadron two days previously. When he re-visited his 'old boat' during its stay on the Thames in 1982, he recalled his 201 days vividly.

"I was allocated to Flight Lieutenant Dave Easton's crew as a very green third dickie. This was to be my crew until September. The day after arriving on the Squadron I was given the opportunity of a trial flight, as an air test was required for R/201. As I came alongside the Sunderland for the first time in the dinghy, it seemed to tower above us.

"It was in fact twice as long, had twice the wingspan, and weighed eight times more than the largest aircraft I had flown previously, an Airspeed Oxford. On the side I saw the squadron letters, NS-R, and the serial number ML814. Little did I think that some 40 years later I would board this same aircraft in the Pool of London by Tower Bridge, accompanied by my 30-year-old son.

"We boarded through the front door into the forward compartment housing the front turret (partly retracted at that stage as the aircraft was at the moorings), bomb aimer's hatch (stowed for the same reason), and the anchor winch. Down each side of the nose were housings for two Browning ·303 fixed machine guns.

"Leading aft on the starboard side was the WC — all mod cons on this aircraft — as crews frequently lived aboard when away from a regular base. In the centre, stairs led to the upper deck; and on the port side a narrow corridor led to the wardroom, with its two bunks and a table in the middle. Then a door into the galley, equipped with a twin primus stove and oven, sink, stowages for crockery etc.

"To either side of the galley were large hatches, which could be opened on the water to stream the drogues when taxiing, thus helping the pilot control the aircraft; and

The flight engineer's position.

(Photo Short Brothers PLC, 79513-3, via Ron Parsons)

also in the air to allow the use of Vickers gas operated machine guns for defence when so required. From here was an additional ladder to the upper deck, this being the access most frequently used by the crew.

"Moving aft again was the bomb room, containing eight 250 pound depth charges, four each side on overhead racks. These racks could pass through the large bomb door to either side and ran out under the wing when in use. This operation was controlled by the pilot; but after the racks had been run inboard again, the hatches had to be closed manually — not always an easy task. It required considerable strength to push the heavy doors up and out into their locked positions.

"Another door with a half partition led to the rear of the hull, with two more bunks, an armourer's bench, and the long walkway down to the rear turret. Also right to the rear was the small sealed hatch in the floor through which the rear-facing camera would be mounted when in flight. This camera was linked to the bomb mechanism and recorded the results of a U-boat attack.

"One of my tasks on this crew was to fix this rear-facing camera in position once we were airborne. In rough weather this was not a job for the squeamish, as back here in the tail the motion of the aircraft could be quite violent.

"If we return to the front entrance and, instead of walking through the lower deck, you squeezed up the forward steps to the bridge, you emerged from under the throttle pedestal, between the pilots. Care had to be taken not to bang your head on the pitch levers

"The 'office' was spacious, and the pilots had good views forwards and sideways. First pilot (captain) was on the port side, with controls for 'George' — the auto pilot — on his left. The second pilot sat to the starboard side, while the third pilot normally stood between the seats. The pilot's panel contained what today would seem very simple instruments, but looked extremely complicated to me then.

"Behind the second pilot's seat was the radar operator in a curtained-off, dark, stuffy compartment — watching a PPI (plan position indicator) roughly eight inches diameter, with its controls to either side.

"Behind this again was the navigator's table, which 40 years on I noted was still there. Behind the first pilot was the wireless op's position, with its MF and HF sets, and the controls for the aerial, which was trailed below the aircraft when in flight.

"All these positions were forward of the main spar. Aft of the main spar on the starboard side, facing the rear, was the flight engineer's position, with the engine instruments, and the fuel tank controls and gauges for the ten wing tanks. Access to the remainder of the upper deck was through the 'oven door', located in the centre of the rear bulkhead.

"Beyond the oven door was the sloping deck, giving access to the mid-upper turret. From this turret there was an extremely good all round view over the wing.

"While I was being shown around prior to the air test, one of the gunners was in the bow adjusting the mooring arrangements, so that we could slip the mooring once the captain was ready. Shortly after all was ready, the outer engines were started, then came the order to the crewman in the bow to slip moorings. The inners were started as we taxied towards the take-off point. At the same time the front turret was wound fully forward and locked into position with five very large screw bolts. Flaps, controls, etc were checked, and the engines tested, while taxiing — then all was ready for take-off.

"All this was somewhat confusing to a 'wheelie' pilot. Even more so when the skipper hauled the stick right back into his stomach and opened up all four engines. The nose came up, and up, and up, with the pilot see-sawing coarsely on the wheel to keep the wings level. Then the stick was eased forward, the nose dropped gently, and we were on the step, accelerating to take-off speed.

"Once in the air, all returned to normal. Although I have done many take-offs since that day, I have never forgotten that first take-off, so outside my experience at that time. At the end of the air test, compared with the take-off, the landing was almost normal. Yet I was still taken by surprise by the way in which the nose rose high into the air as we came down off the step.

"Over the next two-and-a-half months I was to learn that operating flying boats was not the same as land-based flying. For example, an aircraft could not be moved on the water without a pilot being present, not even when being towed by a dinghy or pinnace.

"On land, having parked their aircraft, a flight crew would hand it over to the ground crew and ignore it until required to fly it again. On water, this was not the case. When an aircraft was to be refuelled or bombed up, a skeleton aircrew, including a pilot, must be aboard. This was partly a safety measure, and partly to assist in the particular job on hand.

"Hence, if refuelling was required after landing, a skeleton crew would be detailed off to stay aboard until the refueller arrived. Then, under the watchful eye of the flight engineer, they in fact opened up the tanks and handled the hoses to fill the relevant tanks.

"And, of course, it was always the good old third dickie who fell for the pilot's job on such occasions. In rough or wet weather this was by no means an easy task, although on a warm, sunny day it had its compensations.

The tail walkway. The navigator is checking his D.R. (gyro) compass master unit. Behind him is the workbench. At the rear of the walkway is the entrance to the tail turret.

(Photo Short Brothers PLC, J3 6608, via Ron Parsons)

"On a stormy night, when the 'wheelies' air crews were tucked up safely in bed, the 'web footers' were likely to be called out as storm crew, and have to get out to the aircraft in a dinghy which had little or no protection for either the coxswain or his passengers. Transferring from a bucking dinghy to a rolling, twisting aircraft in pitch darkness — is usually easier said than done. In a violent gale it may even have been necessary to start the engines to ease the strain on the moorings. If these broke, as happened from time to time, then the fun would begin.

"Another marked difference was night flying, because a moving aircraft on water is affected by both wind and tide, and there are no clear taxi tracks to follow to the runway.

"On water, there are not only the lit channel buoys, but also others, the even larger 'battleship' buoys which are unlit. Hence the bowman with his Aldis lamp was an essential aid to the pilot.

"A flare path varied from a line of special pram dinghies, permanently moored in position, to a line of three general purpose pinnaces, lit overall like Christmas trees, and placed as conditions and wind dictated. The result would be, in some cases, a runway with a dog's leg in the middle.

"Height judgement over water in the dark was particularly difficult. A flying boat at night has virtually completed its landing when it is still 200 feet up. From here on, the aircraft continues to descend in the correct attitude and at a constant rate of descent, until it makes firm contact with the water. It will not bounce, provided the approach angle and rate of descent are correct.

"I went on to complete some hundred and fifty hours operational flying, mainly anti-submarine patrols in the Bay of Biscay, and all in NS-R (except for one in her sister boat, ML813, NS-U). Of this flying, one third was flown at night. During this time we saw very little. Perhaps this was not too surprising, as Coastal Command statistics indicated that for each attack made, successful or otherwise, some 1,000 or so hours had been flown."

Unfortunately, it seemed, the pattern was now set for R/201. Her crew took her up for four more Cork patrols during the remainder of June, every one a "without incident". The only very slight change from the monotonous routine was that on the last patrol of the month, 30 June, they left the patrol area 75 miles south west of the Scillies ahead of schedule. They were recalled because of deteriorating weather at base.

If success eluded the crew of R for Robert, they could at least take a little cheer from the overall situation. Those at 19 Group Headquarters could hardly have wished for better. Aircraft continued to make successful strikes against the U-boats, although once all the non-schnorkel boats were eliminated it became a different situation.

In the first three weeks of the invasion, aircraft had sunk nine U-boats and damaged eleven. The tremendous build-up on the Allied beachheads continued to plan, with barely a ripple caused by the U-boats. It wasn't until nine days after the invasion that U-621 became the first U-boat to get an invasion fleet vessel in her periscope sights. It torpedoed and sank an American tank landing ship, but was driven out by Allied naval vessels before claiming further victims.

It was a further two weeks before the next U-boat penetrated the aircraft screen. Dönitz's dream of cutting the Allied supply line and preventing any broadening of the front had been totally shattered, and almost completely by air power. For the admiral who had once said that an aircraft could no more sink a submarine than a crow could attack a mole, it was a bitter pill to swallow.

July 6 saw ML814 and crew on their first sortie for the month. After a tedious afternoon patrolling their box, Cork E, there was sudden excitement just as it got dark. A black object was sighted two and a half miles ahead. The Klaxon sounded and the hull reverberated to the sound of bomb doors clanging down — for the first time ever when on patrol. But the "black object" had by now disappeared without trace — it had all been a false alarm. It was concluded that it had most likely been a whale, and if so was most certainly not the first whale to have quickened the heartbeat of a Sunderland crew.

Looking forward from the tail walkway. The starboard entrance hatch is on the right. The two portholes are in the waist gun hatches (only fitted to ML814 after she was converted to a Mk 5). The ·5 guns can be seen in their stowed positions. The first partition leads into the rear wardroom, and beyond the light coloured bulkhead is the bombroom. On the upper level is the sloping deck, with the oven door leading to the bridge.
(Photo Short Brothers PLC, SU 675A, via Ron Parsons)

Just as well, perhaps, that it wasn't a U-boat. Although the bomb doors had opened properly, the traversing gear had only taken the eight depth charges partly out and then jammed. The flight engineer, Flt Sgt Johns, climbed out into the mainplane trailing edge to attempt to rectify the problem. Twenty minutes later he had the electrically operated mechanism working correctly.

On their next sortie, two days later, it was decided to test the bomb gear en route to the patrol area. The electrical problem was there again. Yet another trip into the port wing for Flt Sgt Johns, but this time a slightly longer one than he had reckoned on. Access to each wing during flight is via a small triangular hatch near the rear of the bridge, one to either side. Wriggling through the hatch is none too easy, even for a smallish and fairly nimble fitter. A while later the captain enquired as to the whereabouts of the engineer, who had not returned to his position. He was nowhere to be seen, certainly not on the bridge where he should have been.

It was only then that the wireless operator recalled closing the hatch some time back because he had felt "a hell of a draught coming in from there". He also remembered hearing a knocking noise from that area recently. The hatch was opened, and out crawled an exhausted and slightly suffocated flight engineer. At least the bomb gear trouble had been rectified!

Four 250 pound depth charges on the starboard side bomb carrier. The bomb door can be seen down (open), to allow the bombs to travel out under the wing for dropping.

(Photo Imperial War Museum, Ref CH 808)

R/201 was called back from that patrol at three the following morning due to weather, and was forced to divert to Mount Batten, touching down in Plymouth Sound at 0630. After their return flight to Pembroke Dock later that day, the crew were due leave. They did not fly Robert again for a month. No one can remember, but it is likely that the boat would have gone up the slip for a spot of maintenance at the same time.

One crew may have been away, but the unrelenting operational effort continued. It was known from the planning stage that Operation Cork wouldn't be a two-day affair, that the patrols would have to continue for months at the least — for so long as the Allies needed to keep supplying men and materials to the continent, and there were U-boats to threaten that supply chain.

The U-boats were still being kept at bay, but the kill rate dropped to almost nil as the non-schnorkel boats disappeared. In those early days of use of the schnorkel, the U-boat commanders were beginning to think that they were almost immune from detection with their new device.

And so ML814 was back on anti-submarine patrol on July 20th, allocated temporarily to another crew under Flying Officer R A N ('Red') McCready. They took R/201 out well beyond the Cork patrol areas, to the south west of Ireland on a CLA (creeping line ahead) search. The use of this type of search method would mean that they were probably following up a reported possible sighting, but it failed to lead them to anything.

The same crew were to make two further sorties in R/201 during late July, each ending with the report "nothing was seen". At the end of one, a night search, they were forced to make use of that very welcome sheltered landing area in the Scilly Isles, as the weather had closed in on all mainland bases.

Second pilot on that sortie was Peter Lillingston, who remembers the night well. "We had a hectic night's flying, there was continual cloud, heavy rain, and fog. The

Scillies looked wonderful in the fresh morning sun, I noted in my diary that they seemed like a vague reef on a silky sea. I believe we came down in St Mary's Sound, mooring in the Main Harbour. There did not seem to be any real facilities there, we were greeted by a Mr Williams who got us some accommodation. We slept under a palm tree in his garden until 4.30 pm, and were served cream buns for tea. It was then time to get under way, and we were back at Pembroke Dock by 7.30, having logged 15·40 hours flying."

Mac's crew, like all others in Coastal Command, faced the problem that all U-boats were now schnorkel-equipped. Crews were straining their eyes to see a mere three to four feet of tube protruding from the surface. The naked eye would be lucky to see this from a mile.

On the odd occasions that the sea was calm, and if the U-boat was travelling fast, then the wake, and possibly a haze of exhaust fumes, were more of a give-away. Even so, five miles was the greatest distance from which a keen eye was likely to see this. But whenever there was any degree of wind, and thus white horses, there was not the slightest chance of seeing a periscope wake.

For the radar operator matters were even worse. Four miles was the greatest distance at which the small echo from a schnorkel could be detected, and then only by a particularly skilled operator and in favourable sea conditions. But as the aircraft flew closer overhead, this small echo would often be lost on the screen amongst the general sea clutter, and this before it was within visual range.

Yet the task was not impossible, as one of 201 Squadron's own Sunderlands had proved. On 11 July Flight Lieutenant Walters and crew had made history in P/201 — the first aircraft ever to locate and destroy a schnorkelling U-boat.

The arrival of August saw Dave Easton's crew back on operations. After an anti U-boat patrol over the Channel in another aircraft, they were back in their boat. It may have felt like old times to be patrolling the Bay in R/201 again, but the optimism with which the crew had gone into the early days of Operation Cork was slowly declining. What chance of success now on a patrol? The U-boat fleet was beaten. The Bay was at peace. Not even the Luftwaffe was to be encountered there these days.

The Germans could of course read the writing on the wall. Allied armies were pushing further into France from their beachheads. Brest was now being approached, and it was only a matter of time before the Biscay ports would fall. The U-boat men were preparing to leave, some of the boats already moving out — back to North German ports, back to Norwegian ports. The tide of U-boats, which had flowed into the Channel from the northern ports as well as from the Biscay ports at the time of the invasion, was now reversed.

Coastal Command had expected a field day as the U-boats were forced out of the Biscay ports, out into the surveillance of the massed air patrols. The schnorkel had dashed their hopes. R/201's crew were told the sad tales in the mess on the evening of the 7th, just back from the first patrol since their break. Not a single crew could remember even sighting a U-boat for weeks.

Two days later there was a buzz around the base, a 461 Squadron Sunderland had actually sighted a U-boat the previous evening. It had made off, but there was no doubt that they had caught sight of one in the glare of their flares. Better news still on the 10th, a 228 Squadron Sunderland carried out a daylight attack on a surfaced U-boat. Was this the great exodus from the French bases? Excitement again at 19 Group Headquarters: they would swamp the Bay with patrols that night. The chances of some kills looked good.

Then came the bad news — Mount Batten was closed by fog. All the Coastal Command airfields in Cornwall were closed by fog, there wasn't a base in southern England which could put up a patrol. So up went the Sunderlands, every boat which Pembroke Dock could get into the air was out. Dave Easton slipped his mooring at 2115, and was on patrol off St Nazaire by just after midnight. This area was to become very familiar to ML814's crew over the coming weeks, as they patrolled almost right into the harbours of the U-boat bases. By one o'clock the following afternoon R/201 was safely back on the trots in Milford Haven — the report again "no incident".

Pembroke Dock's efforts had not been in vain, however, as Ivan Southall from 461 had pulled off a memorable attack that night. After making radar contact he had managed to get his U-boat target silhouetted against the weak moon, and attacked "the old way", no flares, no bombsights. Adding to his worries during the attack was the fact that they were on the very borderline of a prohibited bombing area. He thought it "looked right" as he approached his target, but could it be a British submarine?

As his Sunderland thundered low over the conning tower and he released the depth charges, startled white faces looked up in disbelief. This picture refused to leave his mind for the next hour or so, until at last naval confirmation came through that his victim had been a U-boat.[1]

P/461 had not scored an instant kill, but not far off. After several hours of struggling to control their badly damaged boat, U-385's crew got her to the surface, threw back the hatches, and found themselves surrounded by the vessels of Walker's Second Escort Group. Southall was not disappointed to later learn that his victim's entire crew were safely picked up. Very often a depth charged U-boat suffered heavy casualties and it was not unusual for all hands to be lost.

Three days later another 461 Sunderland made another kill, then on the 18th, No. 201 got their only U-boat for the month. It was Les Baveystock again, in W/201, and his victim was U-107, a boat which had claimed 39 Allied vessels over its career. Baveystock's crew had sighted a wake during a daylight patrol. He turned immediately for the attack, crossing the U-boat just ahead of its periscope, and a mere 50 feet above it. He had released a perfect straddle of six depth charges. "It was too easy", he said later. Perhaps, but it helped morale at 201 no end, and put Baveystock in an elite — those pilots who could claim more than one U-boat sunk.

R/201 and her crew put in four more patrols during August, but still those elusive U-boats evaded her. These sorties were all Rover patrols, taking the Sunderland right up to the French coast off St Nazaire, in the Brest area, and down to the Gironde Estuary. Rover aircraft patrolled not the boxes, but between several fixed points. Their job was to be on hand to fill any break in the chain of patrolling aircraft caused by engine or radar failure, and to investigate further any sighting reports. The effectiveness of the Cork patrol system depended on all the boxes being patrolled all the time.[2]

Several minor incidents did crop up on these Rover patrols, which have remained in crew members' memories over the intervening years, presumably because nothing of real significance did happen.

The first of these occurred on a particularly dark and starless night when R/201 was patrolling in the Ile d'Yeu\Gironde Estuary region. The navigator was seated at his chart table, the chart carefully marked out in green crayon for the daylight work and red for the night hours. With no astro shots possible, he reached up to the

A navigator at work at his chart table.

(Photo Imperial War Museum, Ref CH 420)

Gee box and radio receiver in the hope of getting a fix or bearing, only to find that they were out of Gee range.

Unfortunately, at this moment the engineer decided to pass him a mug of coffee. Inevitably, their arms collided in mid-air, and a goodly amount of coffee finished up on the navigator's chart. Full of apologies, the engineer smartly produced a cloth and proceeded to wipe the coffee from the chart. Before there was time to prevent it, the red night chart work had disappeared. The navigator's ensuing comments are better not repeated, but they were as blue as his small nav. table lamp!

Eventually, a clear enough patch to enable him to get a new star fix appeared, and once the chart had dried out sufficiently he re-plotted the details from his nav. log. Life was back to normality on the bridge, and remained so until 6.45 the following morning when they were back at Pembroke Dock, after a sortie (according to the report) "entirely without incident".

On another Rover patrol, a loose ship's balloon was sighted during the run down to the patrol area. To relieve the monotony it was decided to try to shoot it down, 1300 rounds being expended in so doing. As they flew onwards, the tail gunner watched the balloon slowly descend towards the sea.

Later, when on patrol, they momentarily sighted a yellow dinghy containing three survivors. There were bad sun glare conditions at the time, and they failed to get a second sighting of it, despite a further search. Like all other Coastal Command aircrew, they would do anything possible to assist a ditched crew. They knew only too well that one day it might be themselves down there.

They had heard the stories from those who had come back from such experiences, and thought they could imagine what it would be like. Nights dragging into days until all recollection of time was gone, consciousness passing into sub-consciousness. The sound of an approaching aircraft — did you want it to sight you, or not? Was it ours, or theirs? Had it seen you, would it drop supplies, would it send a rescue craft?

Anxious as they were to re-locate the poor souls in that raft, they just couldn't. Group were notified and other aircraft diverted to continue the search. The crew enquired the following day as to what had been the outcome of this search, and were told that three men had been picked up in the area.

It was about ten in the morning on the last day of August when Dave Easton, his two pilots and navigator, filed into the Pembroke Dock ops. room for their pre-sortie briefing. Much to their surprise, and for the first time in three months, it was not going to be a Cork patrol today. Instead, they were to try to locate the French cruiser *Jeanne d'Arc*, which was steaming northwards, having left Algiers on 27 August. She should have several Royal Canadian Navy destroyers as escort, but the small convoy had been out of contact for several days, and air searches had so far failed to locate the vessels.

As R/201 flew southwards to her search area, some 400 miles south of Fastnet, her crew knew only that they were looking for a convoy including a twin-funnelled French cruiser. In fact the *Jeanne d'Arc* had been detached from the French 3rd Cruiser Division, taking part in the Provence landings, to carry out a special mission. Following the liberation of Paris, members of the Provisional Government, along with officers of the Consultative Assembly, were to be embarked at Algiers for passage to Cherbourg.[3]

At 1705 the convoy was sighted after the radar operator had reported "objects at about eight miles, probably ships." R/201 had found her charges.

The navigator remembers the subsequent events: "The convoy was located and, after exchanging the required recognition signals, and giving the Senior Naval Officer aboard one of the destroyers my calculation of his position, we were requested to carry out a complicated relative search ahead, and keep station with the ships. They were heading for Start Point, and we were with them until last light."

By then the weather was deteriorating, and ML814 was diverted to Calshot. It was 0016 on the morning of 1 September by the time she was safely moored in Southampton Water. Later the same day her crew had the added burden of the return flight to Milford Haven.

Dave Easton's crew were only to make two more patrols in ML814, and curiously, after having served them reliably over all those months, she let them down on their very last flight in her.

They had been on patrol south of Fastnet for about four hours when, at 1530, the starboard outer engine started to play up. After jettisoning his depth charges, Dave Easton set course for base on three engines. Thus there was a certain amount of relief aboard when they saw ahead Milford Haven. In past the old familiar landfall of St Ann's Head came the big Sunderland, and on to a gentle alighting at Pembroke Dock. R for Robert was to be their boat no longer.

This patrol completed a full operational tour for Dave Easton and some other crew members, so the crew was split up. Ron Harris and two other crew members transferred to a new 201 crew being formed.

Dave Easton re-visited ML814 at Calshot in 1992, and recalled those wartime months he had spent in commanding her: "I was proud of my crew, who were loyal, keen, and efficient — Ray Lassiter, spot on with his navigation, Johns the engineer, nursing his engines, and crawling into the wing at times. Chick Mawer, the wireless/radar operator, who picked up just about everything on the water with the very good

ML814 and her first skipper, Flt Lt D Easton, at Pembroke Dock 1944.

(Photo via Dave Easton)

radar we had. Unfortunately, it generally turned out to be a barrel or a bit of wreckage. Off the Brittany Coast there were a lot of poles sticking out of the water, markers of some sort which looked like periscopes, so we had a few false alarms. Weir the fitter was very good for bacon and eggs in the wardroom below. Mooring up at PD was sometimes a trial after a long flight, with tides of 5 knots or more and cross winds, but we usually managed it first time, by slick work with the boat hook in the bow and crafty navigation by me — while we laughed at others going about in circles."

R for Robert was not to leave 201 Squadron just yet, but it was the end of her permanent allocation to one particular crew. There were changes afoot at Pembroke Dock. The Station was going over to Planned Maintenance. This meant that after flying the required number of hours, each aircraft had to go up the slip for certain scheduled maintenance. All very efficient, so the experts said, but added that it would never work if crews expected to have their own boats. And so an excellent system came to an end. The crews were disappointed. Robert had *belonged* to Dave Easton's crew since new, and they had looked after her as if she was their own property. They cleaned not just the perspex and the guns — the necessary items to keep her operational — but the wardroom and the galley as well. They had not turned her into a home sweet home, as some crews did, but she was always spotless, and they had felt proud of her.

It had not only been aircrew, but maintenance crews as well, that had been responsible for their own aircraft. 'Third Dickie' Ron Harris recalls the advantages of a dedicated ground crew: "I remember one time we were plagued by an elusive oil

Return visit. Dave Easton in the captain's seat of ML814 once again, Calshot 1992.

(Photo Peter Smith)

leak on an inner engine. Finally Flt Sgt 'Robbie' Robinson, who was the fitter in charge of the ground crew for R/201, decided that he would have to fly in the wing on an air test. We let down the leading edge platform outboard of the engine concerned, he climbed in, and we closed it up again. We started up, did a quick circuit and landing with him in the wing, then moored up. The leak was thus traced and repaired. Better him than me. I couldn't imagine that happening under planned maintenance. Pride in your own aircraft was no longer there."

Now, life on boats would be that bit nearer to life in the 'ordinary' air force. There would be a list of crews, a list of aircraft, and ops would match the two as suited their needs. So for the rest of ML814's time with 201, which was only to be another four sorties, she would be taken out by different crews.

But things were changing all round at Pembroke Dock. The monthly summary in the 201 Squadron diary at the end of September read: "Although a fairly busy month in respect of number of sorties flown, no enemy aircraft nor U-boats were even seen, let alone attacked. The Bay of Biscay has now been almost cleared of U-boats, and their only activity was one in the region of the Azores and a few in Norwegian waters. They had become exceedingly wary due to their heavy losses throughout the summer. At the end of the month 19 Group, less many of its Squadrons which had been transferred north, was almost quiescent."

The reason, of course, that there were no longer any 'boats in the Bay' was that the Germans no longer held the French ports. Although the U-boats had mostly left the ports by the end of August, the German garrisons fought long and hard before the bases finally fell to the Allies. The US VIII Corps battled doggedly for a month, and suffered 10,000 casualties in the taking of Brest alone.

Life for Sunderland maintenance staff was far removed from that of landplane fitters. The weather here looks reasonable, but the same work had to be done whatever the conditions, if sorties were to be flown.

(Photo Imperial War Museum, Ref CH 859)

Coastal Command intelligence understood that the clearing of the Biscay ports was by no means the end of the U-boat threat. They knew something at least of the development work going on in Germany. For many years experiments had been undertaken with a revolutionary form of propulsion, the Walter Turbine. This required use of the volatile and dangerous fuel, hydrogen peroxide, but would enable submerged operation at high speeds — at last a true submarine.[4]

A new U-boat type, the XXVI, was designed around the new power unit: a much bigger boat of some 2,000 tons, greater range, greater depth capabilities, and a much stronger hull. But developmental problems abounded and, necessity being the mother of invention, a compromise design was arrived at. The new hull was married to a conventional diesel electric power unit with schnorkel. Battery capacity was greatly enlarged, and the new type XXI was born — to be known as the Electro Boat.

U-3501, first of the new type XXIs, had been launched at Danzig in April 1944, and photographed by a reconnaissance Mosquito shortly afterwards. Previous photographs revealed a startling fact, U-3501 had taken a mere six weeks from laying down to launch. How could such a timetable be achieved? Closer examination of the photos showed U-boat sections on adjacent slipways — they were being prefabricated in remote factories and moved to the shipyards for final assembly.

The possible significance of all this was not lost on the staff at Coastal Command Headquarters, and it most certainly meant that the Battle of the Atlantic could yet be lost. To alleviate matters they pressed for, and got, more concentrated bombing of the U-boat production facilities.

With the Bay dead and life at Pembroke Dock apparently running down, were ML814's last few flights with 201 Squadron a guide to things to come?

One of these patrols was over the approaches to the Clyde, and her very last one on 3 October was back out over the open, inhospitable Atlantic. The future for flying boat operations seemed to be either north or west. A considerable part of 461 Squadron moved out of Pembroke Dock in late September for the North Pole, or so the Australians thought. Actually it wasn't quite that far, it was for Sullom Voe in the Shetlands.

Then in October, 201 moved west, back to their previous base at Castle Archdale in Northern Ireland. ML814 did not make the move to Ireland with them however, as at the time she was back at Calshot for a major overhaul.

Footnotes Chapter 2.

[1] *They Shall Not Pass Unseen* — Ivan Southall (p 163)
[2] *Air Power Over Europe 1944-1945* — J Herington (p 144)
[3] *The French Navy In World War 2* — Jacques Mordal (p 332)
[4] *Aircraft Versus Submarines* — Alfred Price (p 180)

See **Appendix 1** for flights made by ML814 with 201 Squadron

Chapter three

Under Canadian Ownership

When 201 Squadron moved out of Pembroke Dock, another squadron, No. 422, moved in. The two squadrons did, in fact, merely swap bases. As 422 came from Castle Archdale, 201 moved to it. No. 422 was to be ML814's next squadron.

While 422 lacked the long history of 201, they appear to have compensated for that through a diversity of experiences in a remarkably short time. The squadron was formed from Canadian personnel during April 1942 at Lough Erne, Northern Ireland (later known as Castle Archdale). It was the fifth RCAF squadron to be formed in Coastal Command during World War II.

The first operational role for 422 was transporting Hurricane spares to Grasnaya in the Soviet Union — support flights for the Soviet Air Force. The Catalinas used for this task also provided cover for the Russian convoys. Next came a spell of ferrying Catalinas across the Atlantic, then back to transport duties — overseas mail down the West African coast to Lagos. It was not until spring 1943 that the squadron started flying regular Coastal Command sorties, and Sunderlands replaced the Catalinas.

When, in late October 1944, word spread through the Squadron concerning the coming move to Pembroke Dock, most people were somewhat puzzled. Why change bases when both squadrons were carrying out the same operations in the same areas? The general consensus was that the move was diplomatic rather than operational. There were two Canadian squadrons at Castle Archdale, 422 and 423, and at times relations became strained between them and the RAF Station administration. Was this an attempt by the powers that be to spread that strain a little?

For whatever reason, the Canadians of 422 Squadron soon found themselves at Pembroke Dock. Just as at Castle Archdale, there were no 'bright lights' nearby, in which to hit the town. They found the watery Welsh beer a poor substitute for Guinness, and there was no substitute at all for the beef-steaks from Ma Bothwell's in Irvinestown. Where else in the British Isles could you get a large steak smothered with two to four fried eggs, and the whole covered with French fried potatoes, all for three and sixpence? Still, at least the permanent accommodation for all ranks at Pembroke Dock was way ahead of the damp Nissen huts at Castle Archdale.

Even in these early days at the new base, there were signs of pending problems between the Squadron and the RAF Commanding Officer of Pembroke Dock, a Group Captain. Basically, it was a problem which was seen frequently in all three services during the war years. Those in for 'the duration' were usually keen to execute the war as quickly and efficiently as possible, and get it out of the way. From time to time such people would find themselves in collision with a pre-war career officer type, to whom the only thing that apparently mattered was that it was all done 'by the book'.

Many a British serviceman found himself irritated by the same problem, but was perhaps more subtle (or intimidated?) in the way he handled it. Units from the Dominions, however, living up to their reputation for colonial forthrightness, were apt to find themselves in confrontation with authority.

Larry Giles' crew at Pembroke Dock. L to R: F/O H G Smiley, F/O T W McNeillie, F/L J C Nesbitt, F/L L E Giles, F/O D A Park, F/O J I Brown, F/L G E Ashie.

(Photo 422 Squadron Association)

One member of 422 Squadron noted in his diary: "The station CO was a tall, dour RAF type who did not impress. Neither did his wife, a grey, up-stage woman whose main interest was the flock of sheep she sponsored to keep the grass down."

It wasn't until 4 December that ML814 was through its maintenance work at Calshot, and flown back to Pembroke Dock by 422's crew 5 under Flight Lieutenant 'Gus' Gauss. Initially, ML814 would be flying on training exercises only. The Squadron had been instructed by 19 Group to concentrate on training and, from their arrival at Pembroke Dock until 13 December, no operational sorties were flown.

Also, the December weather was not proving very conducive to flying but, despite this, ML814 was to put in a number of training flights during the month. Wearing her new squadron letters, DG-N, she took off for the first of these exercises on 8 December, in the hands of Larry Giles' crew. This was a daylight 'Bathmat' exercise of some three-and-a-half hours. Bathmat exercises were primarily radar homing practice. They could be day or night exercises and the target varied. Generally it was a trawler, although dummy schnorkels and Royal Navy submarines were also used. The name derived from Bombing Approach To His Majesty's Armed Trawler.

Although the system of permanent crews for a particular boat was now on the way out at Pembroke Dock, Larry Giles and his crew were to fly N/422 more than any other crew. Two days after the Bathmat exercise they were up in her again on a one-hour low-level bombing exercise.

Then, on 11 December, came a change of crew. David Mills' crew were off on another Bathmat. It may have only been an exercise, but David still remembers it vividly. "This was a quite hair-raising night exercise, testing the new 1·7-inch flare. We tracked an RN trawler somewhere in the stormy and black South Channel, on our Mark III ASV. Then, at a mile distant, 50 feet from the water, and very much on instruments (a good radio altimeter), our brave rear gunner began pushing these

A group at the 422 Squadron Christmas party, Pembroke Dock 1944.

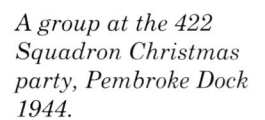

(Photo 422 Squadron Association)

three million candlepower flares through the chute, which electrically ignited the flare. One flare per second was pumped out, each producing an enormous flash, until we pulled up over the terrified target. These pull ups had to be done on instruments to avoid flying into the sea, due to a false horizon cast by the flares. I don't recall using this technique operationally, due to the scaling down of Coastal ops."

Christmas was now approaching, and the Canadians intended to enjoy it. However, some tales were beginning to circulate at Pembroke Dock that these recently arrived Canadians were a bit of a wild bunch. No one in the Squadron could understand how they had ever earned such a reputation, and the new Commanding Officer, Squadron Leader Jack Sumner, was keen to ensure a good name for the squadron. So a Christmas party was organised, and all were to be invited. Besides, it might provide an opportunity to improve relations with the Group Captain.

It was unfortunate, however, that they were not still in Northern Ireland, where many useful contacts had been established. But this was no real handicap to the resourceful Canadians. Dr David Stewart, who was 422's medical officer at the time, recalls the solution: "With Christmas in the offing, the Wingco decided he had business to do in Castle Archdale, and took a group of us over by Sunderland for a couple of days. The most important member of the party was Corporal Johnny LaRue, who had developed a lot of useful contacts in the surrounding countryside. He was, in fact, 422's No 1 'scrounger'. We flew back with a huge number of turkeys and eggs, bags of holly, and a Christmas Tree. We got back just in time to take in the station pantomime by the PD Gremlins — 'Aladdin and his Aldis Lamp'. It was a bit patchy, but the cast were invited to the mess for a drink afterwards and nearly caused a riot, because the RAAF were having a party and didn't like interlopers, however well-intentioned. But it all blew over."

David Stewart recorded Christmas Day in his diary: Dec. 25. "Christmas. Surprisingly, up for breakfast, after cracking a can of the Adj's tomato juice which was most welcome. Nobody showed up on sick parade, so went to church. A 4-star performance, with the station Padre (C of E), the local RC Priest, the SA chap, and our RCAF Padre. After the service we (the Canadians) stayed for Communion. It was

A 422 Squadron crew in a marine tender, waiting to be run out to their Sunderland. For the longer trip down to Angle Bay, an enclosed pinnace or seaplane tender would be used.

(Photo 422 Squadron Association)

a beautiful clear day, with the first hoar-frost we have had so far giving it quite a Christmassy appearance. Then, with the other MOs, visited the local hospital (Woodbine, in Pembroke) and then to the SSQ. (Holylands House, the other side of Pembroke). We all helped serve dinner, but I was the only one who stayed for the meal." Later — "the Christmas dinner in the mess was very crowded; quite good, but I had not completely regained my appetite since noon."

Things were not quite so pleasant that evening when the Group Captain moved in and closed down the dance at the Llanion Barracks early, saying that it was becoming too rowdy. Rumour had it that he also threatened to cancel 422's big party, scheduled for the next day. Wing Commander Sumner was thought to have threatened resignation at this point, so the party went ahead.

Pilot Officer Don Macfie, posted to the Squadron only two days earlier after having completed a tour of duty with 423 Squadron, was at the party. He recorded it thus in his diary: "This evening 422 Squadron had a bang-on party at Llanion Mess hall. A tree, beer, a big singsong — and turkey sandwiches of course. Terry, 'Tuffy' and I were there. W/C Sumner got a good many cheers and is a great favourite with the boys. Even the Group Captain came and made a speech and was cheered. Maybe it was because the boys were pretty merry by this time. It was a good party and came off swell."

Determined as everyone was to enjoy Christmas, there was still a war on. ML814 was out on anti-submarine patrol on 28 December with crew 10 (Larry Giles' crew). In fact, as it had been an 0500 departure, the patrol for the crew had started on the night of the 27th, with all the usual time-consuming preparations for a sortie.

First the ritual pre-ops meal in the aircrew mess — bacon, eggs, and toast. For the pilots and navigator this would usually be rather rushed, as they had to get over for the briefing. What a relief to get into the warmth of the ops room on that clear, cold, and icy December evening. A coal fire was burning as usual in the big, shiny, black stove. The aircrew took in the details marked on the huge wall map of the patrol zone.

Coloured ribbons marked 19 Group's patrols, one of which would be their own. Arrows and numbers showed position and courses for Allied and neutral shipping. As was to be expected these days, only a few markers to show suspected U-boat positions. They listened intently to the long list of do's and don'ts, the hundred-and-one details which the crew, and the captain in particular, were expected to absorb. Then back out into the freezing night. Laden with their charts and other paraphernalia, Larry Giles, Lloyd Smith, Doug Park, and Jim Nesbitt hurried down to the pierhead where they caught up with the rest of the crew, who were even more laden down with provisions and equipment. The eleven of them, wrapped in layers of flying clothing, struggled down the steps and into the pinnace for the long haul down to Angle.

Coming out of the shelter of the pinnace's cabin in Angle Bay, each of them was suddenly struck by how cold it was down here, although unusually there was no wind. Into the planing dinghy they all piled, and before long they were making that final and most difficult transfer, from the dinghy in through the small forward entrance hatch of their Sunderland. Aboard at last, and it already felt as though they had completed a full day's work. What a way to set out in search of the enemy!

Next came all the pre-departure tasks. Two gunners down in the bow struggled with frozen fingers to sort the moorings and get her on short slip. Undoing steel shackles and handling cables and chains in these temperatures was sheer torture. 'Oddie' O'Dell set about his flight engineer's checks. Lloyd Smith prepared his charts up on the bridge. Bilges were checked, signal colours prepared, cameras fitted.

Soon N/422 was airborne, en route for her patrol area in the Channel, between Start Point and the North Brittany coast. Although it is now 47 years ago, Larry Giles still remembers that patrol well: "This particular night wasn't nearly as humdrum as many of our trips flying box patrols. We took off three-and-a-half hours before daylight to do a patrol in the English Channel. Before we arrived there our Gee and radar went U/S, and so of course my navigator wanted to return to base. I decided to stay as most of the patrol would be in the daylight, and we could carry on visually. A while later our radio altimeter started flashing, indicating that we were over land. My second pilot was at the controls, but I grabbed them and made a fast 180 degree turn. We must have been over the Channel Isles, and at that time they were still under German control.

"Shortly after this our radar was repaired so we could now do our patrol properly. On the westward leg of the patrol box my radar operator picked up a U-boat about five miles ahead of us. Before I got within one mile of him he submerged. I carried on with my patrol and went through this same procedure twice more, but couldn't catch him on the surface."

The crew believe that this particular U-boat was later relocated by a Liberator, and successfully attacked. Looking into the records, it would seem almost certain that Larry Giles' U-boat was in fact U-772, a type VIIC, sunk the following day just a little north of where ML814's crew had detected her. The U-boat was operating out of Trondheim, and had sunk a total of five ships. It fell victim to a Leigh Light Wellington of a fellow Canadian unit, 407 Squadron, commanded by Sqn Ldr C W Taylor.[1]

N/422 completed four-and-a-half circuits of its box on this sortie, and was back at Pembroke Dock at 1656. Two days later, 30 December, she was out on her next patrol. An even earlier departure this time, 0150, and a different crew.

George Maier, captain for that flight, remembers the trip not for its excitement, but because of an unusual occurrence: "I recall taking off from PD sometime around

George Maier's crew: Back row (L to R): F/S Sandy Patterson (RAF), F/S Curly Annely (RAF), F/S 'Digger' Grinham (RCAF), F/O Don West (RCAF), Sgt 'Taffy' Rowlands (RAF). Front row: F/O Russ Percy (RCAF), F/L Charlie McCarthy (RCAF), F/L George Maier — Capt (RCAF), 2 Lt Fred McBrien — Nav (USAAF), F/O Ric D'Amico — 2nd Pilot (RCAF).

(Photo via George Maier)

midnight. Our mission was an anti-submarine patrol in the English Channel. Sometime in the mid-morning we picked up a small contact on the radar which caused some excitement for a short period. Then we lost the signal, and spent some time flying around the area trying to regain contact. A naval vessel started signalling us with his Aldis Lamp. Unable to read his message at the distance we were from him, I flew towards the vessel. I had the impression that he had made contact with a sub, and needed our assistance. Imagine our extreme disappointment when, on flying close enough to read the message, it turned out to say 'Happy New Year'. Still, it was well-intentioned."

Back at Pembroke Dock, Wing Commander Sumner wished everyone a happy new year through a message in *Short Slip*, the Squadron newspaper. "In this sixth Christmas at war, we can, for the first time, find reason for the hope that it will be the last Yuletide spent away from our families and homes. My very best wishes to all members of 422 Squadron and 8422 Echelon for a happy and prosperous Xmas and New Year."

David Stewart noted the change of year in his diary: 1 Jan, 1945. "Up for breakfast, a little wobbly but with no definite hangover, which is rather remarkable. I wonder if this year is going to see us home. I hope so, and so does everyone else. It was the most popular of the many toasts drunk last night."

January was to see all flying out of Pembroke Dock severely curtailed by weather, in fact the whole winter of 1944-45 was a particularly harsh one. The Squadron managed to put up a mere 40 operational sorties for the month, and N/422 was only to fly on a few exercises.

On the first of these, a low-level bombing exercise on 7 January, Larry Giles was forced to abandon when his port inner engine failed. Although only the second engine failure on ML814 in almost a year, it was a problem which was troubling 422 a lot, in fact it troubled all the Sunderland squadrons.

A striking view of a 422 Squadron Sunderland taking-off.

(Photo 422 Squadron Association)

When the prototype Sunderland was let down the slipway at Short Brothers' Seaplane Works at Rochester, in October 1937, it was fitted with four Bristol Pegasus engines. Although probably the most suitable British engine available at that time to power the Sunderland, it was already quite a dated design.

That original Sunderland had an empty weight of around 28,000 pounds. With updates in design, increases in armament and other equipment, Sunderlands now had a empty weight of some 35,000 pounds. Taking off for an operational sortie they were usually grossly overloaded, and often at an all up weight of perhaps 65,000 pounds. The aircraft had outgrown its engines. It was necessary to operate them continually at very high power settings, which took its toll on both life-span and reliability.

To compound the problem, when a Sunderland became due for an engine change, the replacement engines were very second-rate. Bomber Command was receiving all the new engines, Coastal Command being given overhauled engines. And these came not from the factory at Bristol, but from the plants of various Midlands car manufacturers. One Sunderland from Number 10 Squadron had seven engines replaced, before it successfully completed one sortie.[2]

There were some frightening stories around concerning engine failures, particularly as the propellers fitted to the Pegasus engine were not of the later fully feathering type. Thus on engine failure the propeller of the dead engine might windmill, and cause additional drag. It was far from certain that a Sunderland would get back to base safely following the failure of just one engine.

Alternatively, overspeeding propellers and shearing reduction gear were also common occurrences. Numerous Sunderland crews had watched in horror as one of their propellers departed. It could take out an adjacent engine as it went, giving them no hope of making base. It could strike the hull with possibly fatal results. Or, and luckily this seemed most common, it could just spin out into the air never to be seen again.

No such disaster struck N/422 on 7 January. It was a straightforward engine failure, and Larry Giles brought her safely back to Pembroke Dock, no doubt relieved that the problem had occurred on a training flight and not far down in the Bay of Biscay, or out over the Atlantic. Yet even the Bay didn't now present the terror that it once had to Coastal Command crews. The Luftwaffe had gone. The biggest dangers now were the weather, and the reliability of those Pegasus engines.

A week later, the weather really struck Pembroke Dock with a vengeance. The morning of 18 January was dark and ominous. Low clouds scurried up the Haven. From the flying control hut on the pier, the duty officer watched the showers approach — each looking like a solid wall of water driving slowly up the harbour

A 422 Squadron Sunderland being brought ashore at Pembroke Dock. With a pinnace on the bow to steady it, a rope has been attached from the tail to a Caterpillar tractor (yellow-nose) on the slipway.

(Photo National Defence Canada)

The tractor takes up the tow. Two men steer the Sunderland by its tail trolley, while two more stand by the main gear wheels with chocks.

(Photo National Defence Canada)

towards him, engulfing everything in its path, then passing. One after another the showers came, and all the time the wind was building.

Even from back in the offices of the solid, stone-walled dockyard buildings, it was now becoming apparent that this was no normal blow. By noon the wind outside was starting to howl. Down on the harbour the marine craft men were worried. It was so rough that all flying, slipway movements, and fuelling had long since been called off.

The biggest concern were the Sunderlands moored down at Angle Bay, as it was far more exposed there than at 'The Dock'. But by now it was too late to do anything about it, too late to tow them home. It would be inviting disaster just to try and get aboard one in these conditions. Normally duty crews would have been put aboard in such weather, but the station had certainly been caught out today.

George Maier has not forgotten that January day: "Our crew was a duty crew that night, but we could not board our designated aircraft due to the very high waves. The idea was that we should run the engines at just sufficient throttle to relieve the strain on the mooring."

Yet it was amazing what crews would do to try and save their boats. A large merchant ship dragged its anchor at Angle Bay, went through the trots, and collected two Sunderlands. Reports from Angle Bay said that there were now three Sunderlands adrift there, all from 461 Squadron, all fully fuelled and carrying depth charges. The Australians put a makeshift crew of ten together to see if they could save any of them.[3]

Just making the trip by pinnace to Angle Bay was a feat in itself. With a dinghy in tow to allow the party to board a Sunderland, the pinnace was often bows under, stern high out of the water, while the tow line snatched viciously. The scratch crew included the squadron CO, its engineering officer, and a senior NCO.

After overcoming various perils, the crew finally got themselves from the bucking dinghy into G/461, which was foundering in shallows but, if left, would be driven onto rocks. They managed to start the Sunderland's engines, got off the shore, and were to spend the next six hours trapped aboard her and fighting the elements.

One of the volunteer crew was the station armament officer, who set about removing the detonators from the depth charges while the flying boat was being thrown violently about in ten foot seas. With Wing Commander Hampshire, the 461 Squadron CO, at the controls, they tried to work their way up to base, but were overtaken by darkness. They put G for George on a mooring, but were torn off it.

Finally, under instructions from an extremely anxious Group Captain, who had been following events from the shore, they ran the Sunderland onto the shoal off Carr Spit. At 2200 hours the ten men came off through the Sunderland's rear hatch and waded safely ashore. The boat was floated off five days later and flew again.

Luckily, 422 Squadron's Sunderlands were mostly on the Pembroke Dock trots on the 18th and ML814, along with the rest of the Squadron's boats, came through the great storm unscathed. Lloyd Detwiller, a 422 Squadron captain, recorded the events in his diary:

"Jan 18 — 120 mph gale blew up. G of 461 broke mooring, beached by crew. 7,000 ton vessel drifted into south trot at Angle Bay. 2 more of 461 Squadron and 3 from 228 Squadron loose. All onto rocks for a total loss. Marine craft in a bad state, but T of 422 was on north trot at Angle and rode out gale OK. Her front bilge had filled, and she was brought back up (to Pembroke Dock) on the morning of the 19th."

The gale passed over but extreme conditions continued. One evening George Maier and crew were due out on an exercise, only to find icing conditions so bad when he commenced his run-up that he called the patrol off. Back at base, he found the Group Captain threatening to Court Martial him, until the 422 Squadron CO intervened to support him. Two Sunderlands from the other squadrons at Pembroke Dock were lost that night shortly after take-off. The following day George received an apology from the Group Captain.

Another story concerning George Maier's crew that winter is worth relating, although the event did not happen when he was flying ML814. It most vividly demonstrates what could be done in a Sunderland under adverse conditions. After a lucky escape one night, landing at Castle Archdale in cloud, George and his navigator worked out a blind landing system. On moving to Pembroke Dock they updated it for use at Angle Bay.

Their system utilised the fact that a Sunderland, once set on the correct rate of descent and speed, will quite smoothly land itself. They selected two approach points on their landing path, one to be passed at 1,000 feet, the other at 500 feet. Each of these points they could locate precisely, by combined use of ASV radar and Gee. They practised the system time and again in good visibility, and felt confident with it.

Arriving overhead one night at Pembroke Dock, they were instructed to proceed to the Scillies — all mainland bases were fogged in. The flight engineer and navigator went into a brief huddle, then George was told that he had nowhere near enough fuel to make the Scillies.

The 'system' was to be tried for real. They came over the 1,000 feet checkpoint on course. The navigator announced the second point, and they were OK at 500 feet. George held course, speed (85kts), and rate of descent (200 fpm), as they approached

A refueller approaching a 422 Squadron Sunderland at Pembroke Dock. A crew member can be seen in the boat's astro hatch, signalling with an Aldis Lamp.

(Photo National Defence Canada)

the hidden waters of Milford Haven. The second pilot, calling off the radio altimeter, said "ten feet", George felt a slight tug on the control column, eased the power, and let the boat come to rest.

Climbing out of his seat, he found the navigator on his knees. "Later that night he told me that he thought there was some higher form of being who did in fact control our lives," George recalls. "This disappointed me, because I thought it was our planning and careful practice which had saved our lives. We were not able to see anything. By shooting off flares, the pinnace finally located us, and we were able to follow it to the mooring area. It gave us a tremendous respect for the Sunderland. In those days it would not have been possible to land in such conditions on a runway with a wheeled aircraft, and so save our lives."

On 24 January, weather again brought the station to a standstill, although this time without disastrous results. It was snowing. David Stewart recalls the day: "We had all been told: 'It never snows in Pembrokeshire'. But it did, one weekday morning, and by the time we went to lunch there was four or five inches of wet snow on the ground. The Canadians then clustered at the mess windows to watch the antics of the Australians, many of whom had never seen snow before. They caught it as it fell, they ate it, rolled in it, and (with a little Canadian coaching) learned how to make snow balls."

It was also his first snow for one member of 422 Squadron, Pilot Officer Straddle. 'Straddle' a Cocker Spaniel, was the squadron mascot. He was looked after by Lloyd Detwiller, who recorded the event in his diary: "Jan 24 — 2-inch snowfall. Big (snowball) fight outside mess. Straddle right in there and enjoyed playing in snow for first time. All ops closed down, whole station in town or out playing in snow."

On the 21 January, Detwiller had been on an operational patrol in ML821 when he was forced by engine problems to make an emergency night landing (without a flarepath) at Mount Batten. Then on the 28th, he flew N/422 to Mount Batten in

Lloyd Detwiller and Straddle.

(Photo 422 Squadron Association)

order to collect his own boat, accompanied by a new posting to the Squadron, Flt Lt Goldthorpe. Presumably Goldthorpe then flew ML814 back to Pembroke Dock. It is not recorded as to whether Straddle made the trip with them, but it would seem likely. Straddle even had his own flying log book, and completed 49 hours in Sunderlands.

But Coastal Command was still fighting a war, even if the load on the squadrons of 19 Group was nowhere near as great as it had been in the months following D-Day. The nature of the U-boat war was ever-changing. U-boat commanders were adjusting their tactics to changing circumstances, and to benefit fully from their new schnorkels.

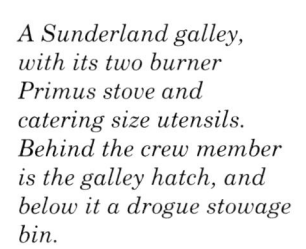

A Sunderland galley, with its two burner Primus stove and catering size utensils. Behind the crew member is the galley hatch, and below it a drogue stowage bin.

(Photo Imperial War Museum, Ref CH 837)

For the first time since the early days of the war, the U-boats moved back to the in-shore waters around the British Isles. Naval vessels had difficulty detecting them by Asdic in shallow water. With the schnorkel they could remain submerged, and so were almost immune to air attack. Merchant vessels plying the coast of Britain were once again targets. One was even sunk within several miles of the Sunderland moorings at Angle Bay!

The answer, obviously, lay in an aircraft which could detect a submerged U-boat. Asdic detected them by transmitting sound 'pings', and finding the bearing of the return echo. To transmit such data from underwater, back up to an aircraft, was asking too much of 1940's electronics. But it was quite feasible to simply listen with a microphone, and transmit back what was heard to a plane.

And so the sono-buoy was born. It was an expendable tubular buoy about three feet long which, after contact with the water, dropped an underwater microphone on a length of cable. What was heard was transmitted back to the aircraft. After about four hours, at the end of its useful life, a soluble plug allowed the buoy to sink. On Sunderlands, they were simply thrown out of the port galley hatch from about fifty feet.

A single underwater microphone was not much use as it could only indicate that there was a submarine in the vicinity. Sono-buoys had to be laid in a pattern, and the technique of their use was not simple. So the crews were packed off to ground school at RAF Limavady in Northern Ireland, then returned and practised the technique. These training exercises, to listen for submarines, were known as 'High Tea' exercises, and ML814 carried out four of them in the first week of February 1945.

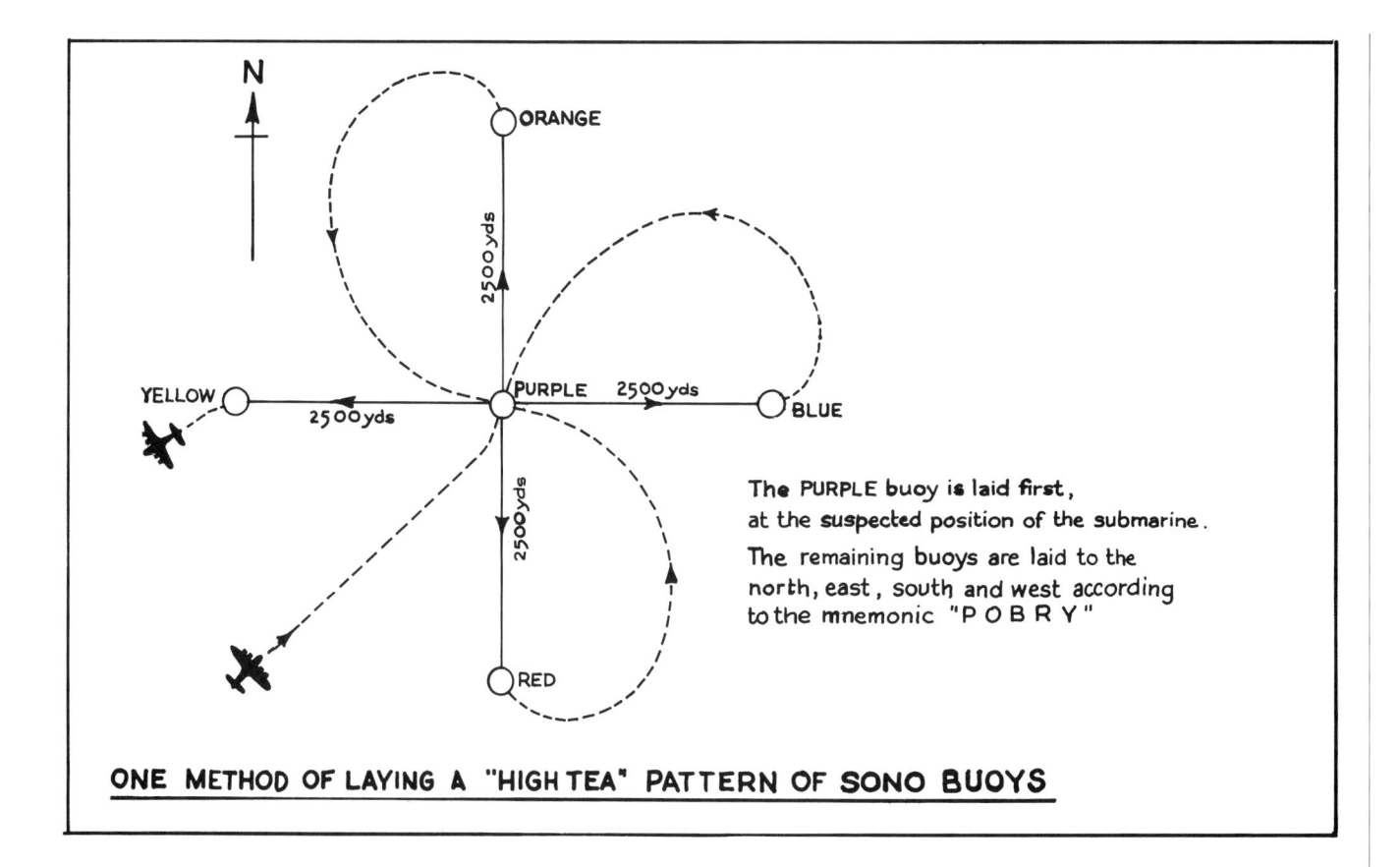

ONE METHOD OF LAYING A "HIGH TEA" PATTERN OF SONO BUOYS

The PURPLE buoy is laid first, at the suspected position of the submarine.

The remaining buoys are laid to the north, east, south and west according to the mnemonic "POBRY"

High Tea — Five sono buoys dropped in a pattern over the area of the suspected U-boat position.

(Drawing Robin Allen)

Flight Lieutenant Ken MacKenzie captained her on one of these exercises, and explains the system: "A single buoy would be dropped over the submerged submarine, or its suspected position. Then we would fly a circular pattern, dropping four more buoys, one at each point of the compass and in a circle around the original buoy. Each buoy had a colour code and transmitted on a different frequency, and the tuner on our radio receiver was colour coded to match. The W/AG (wireless operator/air gunner) had an important role to play in this operation. By tuning to the individual buoys, and listening to relative loudness of the propeller noises, he could determine the course of the submarine. A good operator could even estimate propeller revs, and thus speed of the submarine. If necessary, further High Tea patterns were dropped. The attack would be pressed home or the navy called in to continue the hunt."

This last sentence indicates a growing trend in anti U-boat operations at that time. More and more the aircraft were coming to be used only to locate the U-boats, the actual attacking of them being handed over to the navy. It was an inevitable consequence of the trend to continual submerged operation by the U-boats, but was not welcomed by the airmen.

Larry Giles' crew flew ML814 on two of these High Tea exercises, and the second pilot of that crew was Doug Park. Doug remembers those first High Tea exercises they made, and tells a tale to illustrate just what could be heard from below the surface: "It was later in 1945 and I had my own crew, when one day I sighted a schnorkel just north of the Calf of Man. I went into the attack, and he submerged just as I released eight depth charges. There was no obvious debris so we dropped the

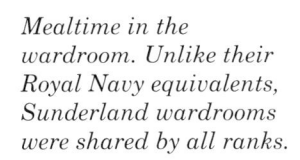

Mealtime in the wardroom. Unlike their Royal Navy equivalents, Sunderland wardrooms were shared by all ranks.

(Photo Imperial War Museum, Ref CH 544)

High Tea pattern and could hear much confusion — sounds of hatches slamming shut and orders being given, before everything went silent. The navy came in and said there was something on the bottom, but in the absence of debris such as the skipper's pants, there could be no confirmation of a sinking."

A week after his High Tea exercises, Doug Park made another flight in ML814: "What will always remain in my memory is that when I was checked out solo in a Sunderland, it was in ML814. This was on February 13, 1945. The day was overcast and there was a light chop on the water in Milford Haven. P/O Hirtle and I were checked out by F/L French (422's Flight Commander) then each did some landings for a total of 1hr 15 mins duration. It was a proud day for me as it was the culmination of my flying to date, about 600 hours on Sunderland aircraft. I went on to a total of around 800 hours on the type, and completed 43 ops trips."

Because of this special link with ML814, Doug Park has kept track of her movements over the intervening years, and still refers to her as the "Pride of my heart". But some of his best memories, as with so many other ex-Sunderland aircrew, centre around the galley.

"Unlike some crews, we had no one specifically assigned to the position of cook. Anyone not engaged on a particular task (such as manning a gun turret) at the time, would fire up the primuses. The standard meal on patrol was steak and eggs, with potatoes, served straight from the frying pan. There was almost always coffee available. This was much to the chagrin of the Liberator, Wimpy and Catalina boys, who had only a thermos and sandwiches."

The galley was even in the mind of 'Bluenose', a regular contributor to the *Short Slip* magazine. Thoughts by now were frequently coming around to how long it would be until everyone could get home. In an article on the subject, he concluded:

Dinner is served. With blackouts on the portholes, the boat is probably on a night op.

(Photo Imperial War Museum, Ref CH 11087)

"However, let us remember that Canada is our home, and let us make Canada proud of us. It will be a proud day when we hear Mother say to Mrs Jones, 'My Bill was overseas. He used to cook in the galley of a Sunderland flying boat!'"

But if the Canadians were wondering how long it would be before they could get home, ML814 wasn't going to be there to see them depart. On 19 February, 1945, she was taken off the squadron's books.

By a curious coincidence, the man who flew ML814 out of 422 Squadron was the same man who had flown her in, two-and-a-half months earlier. On the 20th, 'Gus' Gauss made the two hour flight to Belfast. She was going back to her makers for a refit.

Footnotes Chapter 3.

[1] *Search and Kill* — Norman L R Franks (p122)
[2] *Maritime is Number Ten* — K C Baff (p 335)
[3] *They Shall Not Pass Unseen* — I Southall (p 190)

See **Appendix 2** for flights made by ML814 with 422 Squadron

Chapter four

North and Further North

The reason that ML814, less than a year old, was back at Queen's Island for a major refit, was that a new type of Sunderland had arrived on the scene. This newcomer, the Mk V, was proving so successful in service that it had been decided to return a considerable number of Mk IIIs to the factory for conversion to Mk Vs.

The Mk V was the answer to the Sunderland pilot's prayers. Gone were the old Bristol Pegasus engines, replaced by something more reliable, a little more powerful but, most important of all, with fully-feathering propellers. With the new engines, not only was an engine failure less likely, but if you suffered one it was but a minor hindrance. Even with two engines out, there was a good chance of making base. And there were even the show-offs who flew on one engine (but not for long).

What was this remarkable new engine? It was the American Pratt and Whitney Twin Wasp, although it was far from new, having first gone into production back in 1930. Many Sunderland pilots had also flown Catalinas, powered by two Twin Wasps, and had made the obvious suggestion — why not put them in Sunderlands?

But it had generally been believed that the engines were too heavy and powerful for the Sunderland wing. Then the innovative No. 10 Squadron at Mount Batten started to look at the problem seriously. After discussions with Shorts design staff at Rochester, it was found that the wing could in fact take the bigger engines. Two prototypes were built, one by Shorts, one in the 10 Squadron workshops at Plymouth. Tests showed them to be so successful that the Mk V was born.

Although the engines were the main change, there were others. Now that the Germans were out of France, Luftwaffe fighters were no longer a threat. So the mid-upper turret, which created a lot of extra drag, was deleted. Two new guns were added to partially cover the arc of fire of the lost turret. These were the waist guns, mounted by new hatches just aft of the wing trailing edge. In fact they originated from another No. 10 Squadron modification, copied from the installations on various American bombers.

When ML814 came out of Shorts at Belfast, in late April 1945, given a new lease of life as a Sunderland Mk V, she was allocated to No. 330 Squadron. Like 422, 330 Squadron was a 'new' squadron, a temporary wartime unit.

The Squadron had been formed in Iceland during April 1941, from Norwegian personnel who had escaped when Germany overran their country. They were actually drawn from the Norwegian Navy, as the country did not at that stage have a separate air force. Throughout the war, they proudly retained their naval uniform and ranks, rejecting any attempts to turn them into just another part of the RAF. Besides, their own uniforms were far more suited to their work, and the climate in which they worked.

At first they were equipped with Northrop N3P-B floatplanes, purchased by the Norwegian Government (in exile), and commenced operations with 18 Group Coastal Command. Later, Catalinas were added to the Northrops, greatly extending the range of operations. Early in 1943, the Squadron moved to Oban, in Scotland, where it re-equipped with Sunderlands. In July it moved again to the flying boat base

ML814 in service with 330 Squadron.

(Photo 330 Squadron Royal Norwegian Air Force)

of RAF Sullom Voe, located on an inlet of the Shetland Islands and about 25 miles north of Lerwick.

There was no Coastal Command station which could even approach Sullom Voe for remoteness and for bleakness. It was on the same latitude as that of southern Greenland. A crew flying into the base for the first time was usually aghast at the desolate, colourless landscape, devoid even of a single tree. Nothing but a grey foreground stretching away to dark hills, the whole surrounded by water.

The base itself was a dispersed collection of black huts at the head of Garth's Voe, a small bay off the long, deep inlet of Sullom Voe. When a Sunderland alighted, and its crew threw open the hatches to taxi in, it seemed as if they were in a different world. The air had a freshness such as they had never smelt in mainland Britain. When they picked up the mooring, they found the water so clear that they could gaze from the buoy down the full length of the mooring chain to the sea floor.

Sullom Voe was base at different times during the war to seven squadrons of Coastal Command, but the Norwegians were probably the only ones to look on it as home. It was, after all, as close to their homeland as they could possibly be, while still on British territory. Whether they realised it or not, they had made something of a legend for themselves in Coastal Command. Ivan Southall, an Australian Sunderland pilot, wrote of them: "The Norwegians flew from Sullom Voe. They were big blond fellows who flew the big way they looked. They took their Sunderlands even into the fjords, even to shoot up German outposts. That was their pleasure and they were welcome to it. They flew in the climate they had been bred in. They weren't chilled by the gales or disheartened by the mists. They flew those seas because they loved them." [1]

When ML814 arrived at Sullom Voe, things had not been going well for the Norwegians of late. Early April had brought hail and snow showers, certainly not unknown weather in these parts, but unseasonably bad. The 5th saw the loss of a patrolling Sunderland and its entire crew, when it caught fire and then attempted an unsuccessful landing in the open sea.

Captain Evensen's crew at Sullom Voe. Back Row: (L to R) Thomassen (WOP), Karlsen (A/G), Markant (WOP/AG), Holdsvik (engineer), Aåsland (chief engineer), Markussen (A/G) Front Row: Haakonsen (A/G), Evensen (capt), Ringe (navigator), Vold (2nd pilot).

(Photo via O G Evensen)

The Squadron had been having more than its share of problems with unreliable Pegasus engines. In one month alone its maintenance staff had made 13 unscheduled engine changes. These problems were compounded at Sullom Voe by poor facilities, and by the months it sometimes took to obtain spares from the mainland; not to mention the sub-arctic weather conditions in which maintenance crews had to work.

So the arrival of nine new Mk Vs on the squadron, with their Pratt and Whitney engines, was the most welcome gift imaginable. ML814 now carried a new set of squadron letters, WH-A, plus the regular 330 Squadron embellishment — a small Norwegian flag painted under the pilot's side window (it was small only because the RAF wouldn't allow it any bigger).

On 2 May, Captain Georg Evensen, with second pilot Qm Nesheim, navigator Second Lieutenant Ringe, and eight other 'airborne sailors' (as they liked to refer to themselves), took A/330 on her first anti-submarine sweep with the new squadron. As was frequently the case with operations from Sullom Voe, this patrol was providing a screen for one of the Russian convoys, today convoy JW66. The patrol was a typical one — "low cloud, some icing, sea moderate. Many fishing boats seen and investigated. Nothing of significance to report."

Two days later, Second Lieutenant Torgersen's crew were off in her on a similar patrol, recorded in the operations record book as follows: "Up 1320. Down 0105. Patrolled box 3 times. 2043 Over sailing vessel *Haraldur*. 2357 Off patrol. Several fishing vessels and friendly aircraft sighted during patrol. Average height 800 ft. Sea moderate. Darkness 2330-0105. Radar U/S 1515-0105."

Over the latter part of 1944, and the early months of 1945, the war against the U-boats had reached something of a stalemate. The U-boats were not sinking an inordinate tonnage of Allied merchant vessels, the Allies were not sinking many U-boats. Dönitz considered that the main value of his fleet was in keeping a huge number of Allied aircraft and naval vessels 'tied up' in controlling the U-boat threat.

Meanwhile, his hopes were still pinned to the new types of U-boat. At the beginning of 1945 the German Navy had 51 of the new type XXI U-boats in commission, working-up in the Baltic. Yet even here, in what they had considered was their private lake, the U-boats were badly harassed — mainly by air-dropped mines. When U-2511 nosed out of Bergen on 30 April 1945, she became the first type XXI to set out on an operational cruise.

As the big new craft was passing submerged through the North Sea, she encountered a Royal Navy submarine hunting group. U-2511's commander simply retracted his schnorkel, went deeper, and increased speed to 16 knots. The warships evaded, he continued towards the Atlantic.[2]

At last Dönitz had what he wanted, a U-boat which, in sufficient numbers, could again massacre the Allied convoys. But was there now time? Russian forces were advancing into Germany from the east, forcing the Baltic naval bases to be abandoned one by one. The U-boats, over 70 of them, headed west for the bases in Norway. But to get there they had to pass through the narrow Kattegat which was both shallow and mined.

This forced the U-boats to the surface, and Coastal Command's strike aircraft — rocket-firing Mosquitos and Beaufighters — were expecting them. From the start of April until the end of the war, 26 U-boats were lost in this area. It was Coastal Command's 'happy time'.

But while all this was happening, the Sunderlands from Sullom Voe were still out patrolling further north, their crews still risking their lives. On 7 May, ML814 was out yet again, this time under Captain Bugge, and again in the area to the north west of the Shetlands. It was to be yet another "Nil to report" patrol.

As Captain Bugge's crew kept up their tedious, mesmerising search for something as near invisible as a schnorkel wake, another Sullom Voe aircraft patrolling nearby had more luck. A Catalina from 210 Squadron depth-charged the swirl left as a schnorkel head disappeared rapidly. Not seeing any result from the attack, they dropped a High Tea pattern.

The sono-buoys transmitted noises characteristic of a submarine stopped in the water, after which intermittent engine and machinery noises were heard. It was later discovered that she had foundered with all hands. The 210 crew did not know it at the time, but this was to be the final U-boat to be sunk by Coastal Command in the war.[3]

For the Norwegians at Sullom Voe, a far more important event than the mere sinking of a U-boat was taking place that day. At 1700 hours, just as Captain Bugge was touching down on the Voe in A; Captain Björnebye in Sunderland G of 330 Squadron was flying up Oslo Fjord at an altitude of 1000 feet, firing off red flares.

Aboard G/330 were ten military VIPs, flying in to accept the surrender of Norway from General Böhme, the Supreme Commander of all German forces in Norway. When Captain Björnebye finally arrived back at Sullom Voe on the completion of this mission, it was the evening of 10 May — by which time full German surrender and

Surrendered U-Boats at Lisahally, Northern Ireland.

(Photo George Maier)

the end of the European war had taken place. At midnight on the 8th, Grossadmiral Dönitz made the formal announcement of German surrender. Following Hitler's death in his bunker, Dönitz had become leader of the Third Reich. On the U-boat frequencies, messages were transmitted giving instructions to their commanders to surface, broadcast their number and position, and proceed on the surface to certain designated ports.

They had to fly a blue or black flag, and were often formed into small convoys and given naval or air escorts. For many Coastal Command crews it was their first chance to get a good look at a U-boat. It was the end of the greatest anti-submarine battle ever, the outcome of which was always the most critical factor in determining Britain's ability to survive the war.

The German U-boats had accounted for more than 2,500 Allied merchant ships, totalling around 14 million tons. Out of 1,160 U-boats commissioned, about 730 were destroyed by enemy action. Of these, 288 were sunk in open water by aircraft operating alone, and two thirds of these by Coastal Command.

Yet, despite all those many hours that ML814 spent on patrol, and all those skilled aircrew of various nationalities who manned her, she never had the opportunity to attack a single U-boat. Sad, perhaps, that when coming to write the history of this one flying boat, there isn't a victory or two to notch up to her credit. On the other hand, it was very typical. For every aircraft taking part in the Battle of the Atlantic which scored a U-boat kill, there were very many which did not.

The war may have officially ended, but it was not over yet for Coastal Command — not quite. Although the U-boats had been ordered to surrender, there were strong doubts as to whether some of the commanders would actually do so; they may try to escape to neutral countries, or may even indulge in piracy. And so, almost until the end of May for some squadrons, the anti-submarine patrols continued.

Many U-boats did surrender, although some were a little tardy in doing so. A few made their escape to neutral ports, even as remote as Argentina, not that it did them

Crown Prince Olav of Norway (in British battledress) takes the salute on arrival at Oslo, 15 May 1945. ML814 had flown escort to his convoy.

(Photo Imperial War Museum, Ref BU 6167)

any good. But a large number of U-boat commanders, incredulous that their own superiors were ordering them to hand over their craft to the enemy, simply scuttled their boats — over 200 in the North Sea alone. [4]

While other Sunderlands were still out on their patrols, ML814 had a special mission on 11 May. She was to escort a convoy which was departing from the naval base at Rosyth in Scotland that morning. The convoy consisted of HMS *Devonshire*, two Manxman type cruisers (*Ariadne* and *Apollo*), and 3 destroyers (including HNMS *Stord*), all making the two-day voyage to Oslo. Aboard the convoy were General Sir Andrew Thorne (Commander, Allied Liberation Forces Norway), Crown Prince Olav, and Ministers of the Norwegian Government. King Haakon was not in the party, as it was considered that the situation in the country was too dangerous as yet.

Captain Evensen had the honour of this escort duty, and noted in his log that also flying escort for part of the time were six Mosquitos and six Beaufighters. The Sunderland sighted a mine in the path of the convoy, and guided in one of the destroyers to deal with it. During the flight, A/330 developed a fire in her flap-operating motor, which was safely extinguished.

The officers photographed at Sullom Voe in May 1945, to celebrate the European ceasefire. The Norwegians are easily distinguished by their naval uniforms. Second row centre is Grp Capt C H Cahill, DFC, AFC, (CO Sullom Voe). On his left is Cdr C R Kaldager (CO 330 Sqn), on his right is Wg Cdr R W Whittome (CO 210 Sqn).

(Photo via Maj Gen C R Kaldager)

Ahead of General Thorne lay a difficult task. In the latter years of the war, he had been General Officer Commanding-in-Chief, Scottish Command. As such, he had been responsible for liaising with and assisting the Norwegian underground, and planning for the invasion of Norway should the opportunity arise. In fact, many of the Coastal Command anti U-boat patrols out of Sullom Voe had been planned on the basis of information provided by the Norwegian underground.

Now General Thorne had to establish order in Norway, where there were still some 400,000 fully-armed German troops. They were, in their own minds, still undefeated, as they had not come up against any Allied military forces (except for the 20th Mountain Army in the far north, who had been beaten back by the Russians). How they would react to the situation, and how General Thorne could control them with the relatively small number of Allied troops at his disposal, was yet to be seen. He could, however, count on the support of the Milorg, the well organised Norwegian underground movement.

From Sullom Voe, 330 Squadron was still patrolling. On 12 May they lost yet another Sunderland, forced by engine trouble to make an emergency alighting on those harsh northern seas, 60 miles from the Faeroes. The crew were rescued, but the boat itself lost during an attempted tow. The human enemy might be beaten, but the weather, combined with mechanical failure, had always claimed more Sunderlands than the U-boats.

With this event fresh in their minds, it was with considerable relief that the 330 Squadron crews learned on 14 May that they were being taken off-line from operational patrols. At the same time, they discovered that there was still much work ahead for themselves and their Sunderlands, back home in Norway.

Down south, most of the temporary wartime squadrons would shortly be wound down. At Pembroke Dock, 422 Squadron prepared to move to the Pacific as a transport squadron with Liberators, but later this plan was abandoned and 422 was disbanded. The Australian squadron at Pembroke Dock, No. 461, was disbanded in June. For 201 it was different, they were a permanent squadron, and eventually they moved back to their traditional home, Calshot.

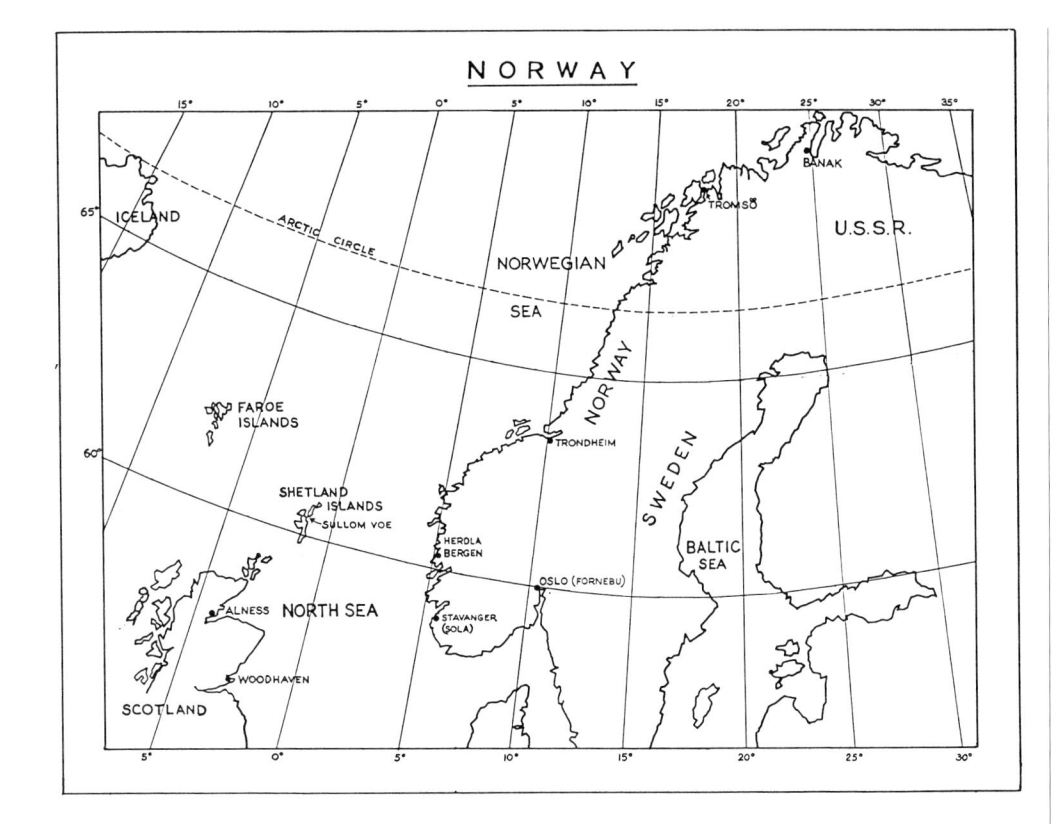

Map showing principal ports of call for ML814 with 330 Squadron.

(Map by Robin Allen)

But ML814's new squadron, No. 330, was given an extended life. They were to relocate to Norway, to assist with General Thorne's task of returning the country to normality. Their new base was to be Sola, near Stavanger. Pre-war, Sola had been a Norwegian Navy flying station, and of course it was taken over by the German forces during the war. There were both flying boat facilities and an adjacent airfield.

The Squadron's Sunderlands started moving personnel and gear to Sola on 14 May, and there were also many transport flights from Sullom Voe to Woodhaven, a small base opposite Dundee on the Tay. Throughout the war, Woodhaven had been base to several Catalinas of 333 (Norwegian) Squadron, involved largely on 'special' flights — taking agents into, and rescuing them from the Norwegian coast. Now much RAF equipment and personnel was being returned to here from Sullom Voe.[5]

A/330 set out for a transport flight to Woodhaven on 25 May, but was forced to return to Sullom Voe with engine trouble. It would seem that this trouble kept her grounded for some time. Meanwhile, the other Sunderlands of the squadron remained busy on transport duties, including the shuttling of military personnel between Woodhaven and Norway. On 13 June, A/330 made a return trip from Sullom Voe to Sola, taking nine of the squadron's personnel plus some equipment over, and then on the following day she returned to Sola permanently.

By then, the Sunderlands were already accepted as a vital part of the Norwegian transport network, in the difficult days following the German surrender. Efficient transport was essential to General Thorne's work of controlling the German troops and rehabilitating the country. The Norwegian transport infrastructure had been badly damaged, and the country's pre-war airline had ceased to exist.

General Thorne's first task was to round up all quislings, then to neutralise and repatriate 400,000 German troops. Vidkun Quisling was the Nazi sympathiser who had been puppet ruler of the country during the war. General Thorne moved into Quisling's ostentatious Oslo mansion for his six-month stay in the country.

As a form of stop-gap transport, the nine Sunderlands of 330 Squadron were ideally suited to the needs of the time. There was no other available aircraft with anything like the carrying capacity of a Sunderland. And the Norwegian coastline, with its fjords offering sheltered landing areas, was ideally suited to a Sunderland's needs.

And so, from May until November 1945, 330 Squadron was to operate an airline. Like any other airline, there were scheduled daily flights to a timetable, between scheduled ports of call. Like any other airline, it was fairly routine, even tedious, work. On the other hand, operating in and out of those fjords in all weathers, without the usual airport navigational aids, could at times test the crews' skills to the limit. The ASV radar proved to be even more useful for navigation than for finding U-boats.

The routes operated by the Squadron linked their base, near Stavanger, to the capital, Oslo; and to the coastal towns of Bergen, Trondheim, and Tromsø. Oslo was an hour-and-a-half's flying time from Sola, and its airport, Fornebu, handled both marine and land aircraft. There was a daily service from Sola to Bergen (50 minutes), and from Bergen on to Trondheim (2 hours 15 minutes). The furthest point north regularly served was Tromsø, which was four-and-a-half hours flying time beyond Trondheim, and within the Arctic Circle.

On 20 June, ML814 began flying the 'bus service'. That day she flew across to Fornebu, and from there north to Trondheim. On the 21st she made the return from Trondheim to Sola. On the 22nd she took the midday flight to Bergen, arriving at 1300. The return departed at 1700, back at Sola 1800. Same timetable and destination on the 23rd. And again on the 24th. So life was to continue for ML814 and her crews, and for the other Sunderlands of 330 Squadron, for many months to come.

Second Lieutenant Vold was captain when, on the last day of June, ML814 made her first flight to Tromsø, capital of the Norwegian Arctic. No problem with night flying here at this time of year, it was perpetual daylight! It would seem likely that on these flights she flew as far north as any Sunderland had been up until that time, although this was later to be eclipsed when a number of Sunderlands air-lifted a British expedition onto a lake in North Greenland.

Tromsø was the gateway to the nearby province of Finnmark, which needed to be completely rehabilitated after the war. Most of the flights to here were supporting this work, and amongst the passengers on this trip had been 14 nurses. When the liberating Russian forces had moved into Finnmark in October 1944, the Germans had retreated to Tromsø, carrying out a devastatingly thorough scorched earth policy as they did so.

Tromsø itself is on a hump-backed island, the adjacent mainland and surrounding islands all being mountainous, and ice-covered even in midsummer. Thankfully the town, with many nineteenth century wooden buildings, had escaped the destruction of the previous year. It was from here that King Haakon and his Government had left for Britain, on HMS *Devonshire*, in June 1940.

To approach Tromsø by Sunderland, normally through a mountain pass, and then to touch down between the islands — surrounded by such scenic grandeur — was a memorable experience for all those who visited the Arctic this way. On their flight northwards, the Sunderlands often passed over the upturned hull of the *Tirpitz*.

General Thorne travelled widely throughout Norway, checking on the progress of his forces in collecting the Germans into special 'reservations', and encouraging the recruiting of Norwegian Army units, which it was hoped would allow the release of Allied troops from the country. And so on 12 July, ML814 (commanded by Second Lieutenant Lorentzen), had the task of flying him from Sola to Fornebu.

The next day, the General left for a meeting with General Eisenhower, who invested him with the United States Legion of Merit. The Norwegian situation was discussed, and it was agreed to aim at having all foreigners out of Norway by the end of November. Although the German forces were no longer a danger, progress on their repatriation was going to be slow. In addition to the troops, there were around 100,000 displaced foreign nationals in the country yet to be relocated.

All of this meant no let-up in the continuing transport duties of 330 Squadron. Operating in such circumstances often produced unexpected problems. On 18 July, Captain Evensen left Sola for Tromsø, although he was delayed on the flight north by bad weather. The next day he flew further north to Banak, where, alighting on Billefjord, he found that he had to touch down in between two minefields. On the return flight he had 22 passengers for another unusual destination, Herdla, just north of Bergen.

Despite the fact that the 330 Squadron Sunderlands were operating daily with a considerable number of passengers aboard, no special arrangements were made for their accommodation. The Sunderlands allocated to BOAC for transport duties were

A 330 Squadron Sunderland at Fornebu (Oslo), in May 1945.

(Photo Norwegian Defence Museum)

at least fitted with rudimentary seating. But here in Norway, the passengers had to make do with what was available — the few seats in the wardroom, the bunks, or sit on their own baggage.

There were no heavy maintenance facilities at Sola, at least not for Sunderlands, so when a check became due this necessitated a trip south to Calshot. Thus on 23 July, ML814 made the seven hour flight to Southampton Water. It was 10 August when she returned to airline duties at Sola, and shortly afterwards found herself making two more round trips to Calshot — one with 30 passengers, one with 5,000 pounds of freight.

Right through August, September, and October the 'bus service' duties continued without let-up, and gradually a degree of normality was returning to life in Norway. General Thorne was utilising a somewhat audacious, but successful, method of dealing with the vast number of German service personnel. He retained and respected the German command structure, even to the extent of officers still carrying their pistols.

He utilised the Germans' own infrastructure not only to look after their own needs and eventually to transport themselves home, but also to carry out reconstruction work. In conjunction with this, several Dornier flying boats were used on mine clearance and transport, particularly ambulance work — still manned by their German crews.

Another curious tale involving 330 Squadron at Sola concerned an Arado 196A floatplane. When the Squadron arrived at the base, they discovered this brand-new German aircraft. It was quickly painted into Squadron markings and put to use as a communications plane, proving extremely popular with the pilots. But it was threatened by an aspect of Allied policy which caused considerable friction between the Allied Command and the Norwegians.[6]

It was Allied policy that all German war material not usable for civil purposes should be destroyed. To the Norwegians, who had lost most of their own equipment, and even personal possessions to the Germans, this was wanton destruction. The personnel of 330 Squadron, trying to avoid the confiscation of their Arado, entered its serial number as that of a Swedish-owned civilian Arado. Despite their protests and subterfuge, it was eventually taken by the Disarmament Wing and blown up.

Early in November the Squadron lost a Sunderland, wrecked at Trondheim after being blown from its overnight mooring during a storm. On the 8th, ML814 arrived back at Sola from another maintenance visit to Calshot. The end of operations for the Sunderlands was by now in sight. Norway's pre-war airline, DNL, was being been re-formed, and a civil infrastructure was developing. The 21 November saw the transfer of 330 Squadron from the Royal Air Force to the new Royal Norwegian Air Force. The sailors were now officially airmen. But to keep the Sunderlands, it was decided, would be beyond the finances of the new air force, and so they would be returned to Britain.

British crews came up to Sola in December to collect the boats. One pilot amongst them was Ron Harris, who recognised ML814 as his old friend R for Robert of 201 Squadron days. But the crew due to fly ML814 home were in for a surprise. She had suffered damage to her fin, and a new one would have to be fitted. It was en route overland from Oslo, but marooned by winter snows.

Beginning its pattern of always being the last, ML814 was not to leave Norway until late January 1946. A crew arrived from Alness in Scotland and made two test

Another shot of ML814 while in service with 330 Squadron.

(Photo 330 Squadron Royal Norwegian Air Force)

flights, on the 28th and 29th. On the 30th they departed for Scotland. Ron Richmond was second pilot on the flight, and has certainly not forgotten the day: "Thirty minutes out from Sola we were having icing problems, and had to return, discovering that our de-icing tank was empty. The Norwegian Air Force ground staff had drained the de-icing fluid to drink. Stavanger was a 'dry' area at the time."

Finally, on 31 January, they successfully made the return to Alness, flying in typical North Sea winter weather — "as best we could at approximately 200 feet, to stay under the cloud base." It was the end of an unusual chapter in the life of a military flying boat, but probably also the end of her active life. There was no requirement for additional Sunderlands in the Royal Air Force of 1946.

Although all the Sunderlands had now departed from Norwegian waters, it was not to be the end of big flying boats in the fjords. Such was the impression made by the Sunderlands, that when DNL started re-equipping, they bought a number of Short Sandringhams to fly the same routes as had their predecessors. The Sandringham was simply a civil airliner built by converting ex-RAF Sunderlands.

As a final footnote to the Norwegian story, 330 Squadron (Norwegian now, of course, and not RAF) still flies out of Sola. They provided considerable assistance with the compilation of this chapter. Sadly (for the romantics), Sea King helicopters have replaced the flying boats.

Footnotes Chapter 4.

[1] *They Shall Not Pass Unseen* — Ivan Southall (p 181)
[2] *Aircraft Versus Submarines* — Alfred Price (p 226)
[3] *ibid* (p 225)
[4] *Maritime Is Number Ten* — K C Baff (p 392)
[5] *Flying Cats* — A Hendrie (p 46)
[6] Typewritten report by General Sir Andrew Thorne, held at Imperial War Museum, London.

See **Appendix 3** for flights made by ML814 with 330 Squadron

Chapter five

Pacific Days

Alness, near Invergordon on the Cromarty Firth, was only a port of call for ML814. Like many of the other wartime flying boat bases, it would soon be closing down. From Alness she was flown to Killadeas in Northern Ireland.

Killadeas was on Lough Erne, just along from Castle Archdale. A flying boat base had been built there for the US Navy, but with the outbreak of the Pacific War they never utilised it, and the RAF took it over as a training base. ML814 was brought here for storage, being entered on the books of 272 MU on 6 February 1946.

Before the year was out Killadeas was due to join the rapidly growing list of ex-flying boat bases. ML814 was again relocated, this time returning to Wig Bay in Scotland. Here, on 8 November 1946, she was put into the care of 57 MU, the unit who had prepared her for going to war when she was a brand new Sunderland, back in early 1944. Little likelihood now of an exciting future ahead. At least she had escaped the fate of many Sunderlands at the war's end — either cut up for scrap or simply towed out into deep water and scuttled. This was the usual fate of the earlier marks.

Being a Mk V, it was considered worthwhile mothballing ML814. She was picketed down outside, amongst rows of other Sunderlands. Removeable items such as guns and radios were taken out, covers fitted over gun turrets, cockpit windows, and anywhere else likely to let in the weather. The engines were inhibited to prevent them from deteriorating, then covers fitted over them as well.

From time to time, Sunderlands in the few remaining operational squadrons were written-off, and a boat would be retrieved from Wig Bay as a replacement. A few were re-purchased by Shorts, taken back to Belfast, and converted into Sandringhams.

Six years later, in 1952, an order was placed with Shorts for some refurbished Sunderlands. Since the war, batches of Sunderlands had been sold to the French Navy and the South African Air Force, but this latest order came from New Zealand. The Royal New Zealand Air Force wanted 16 boats, which were to cost NZ£1,050,000.

Such an order caused considerable surprise. It was now 15 years since Sunderlands had first entered service, and in most countries flying boats were being replaced by land planes — even in the maritime reconnaissance role. Coastal Command had now been operating its new Shackletons for over a year, and it was apparent that the days of Sunderlands in the RAF were numbered.

But the New Zealanders knew what they were buying. During the war they had operated four Sunderlands as transports in the Pacific. With these and some Catalinas, they virtually ran a transport airline for the Americans. After the war the Sunderlands went into civil use, and by now the Catalinas had come to the end of their useful life. There was no aircraft more suitable for New Zealand's wide Pacific domain than the Sunderland.

The batch of sixteen was selected from those at Wig Bay, and ML814 was amongst them. In May 1952 she was pulled out of mothballs and flown back to Queen's Island, Belfast where, along with the other fifteen, she underwent a total refurbishment. The old boat was being given her second new lease of life.

THE SOUTH PACIFIC

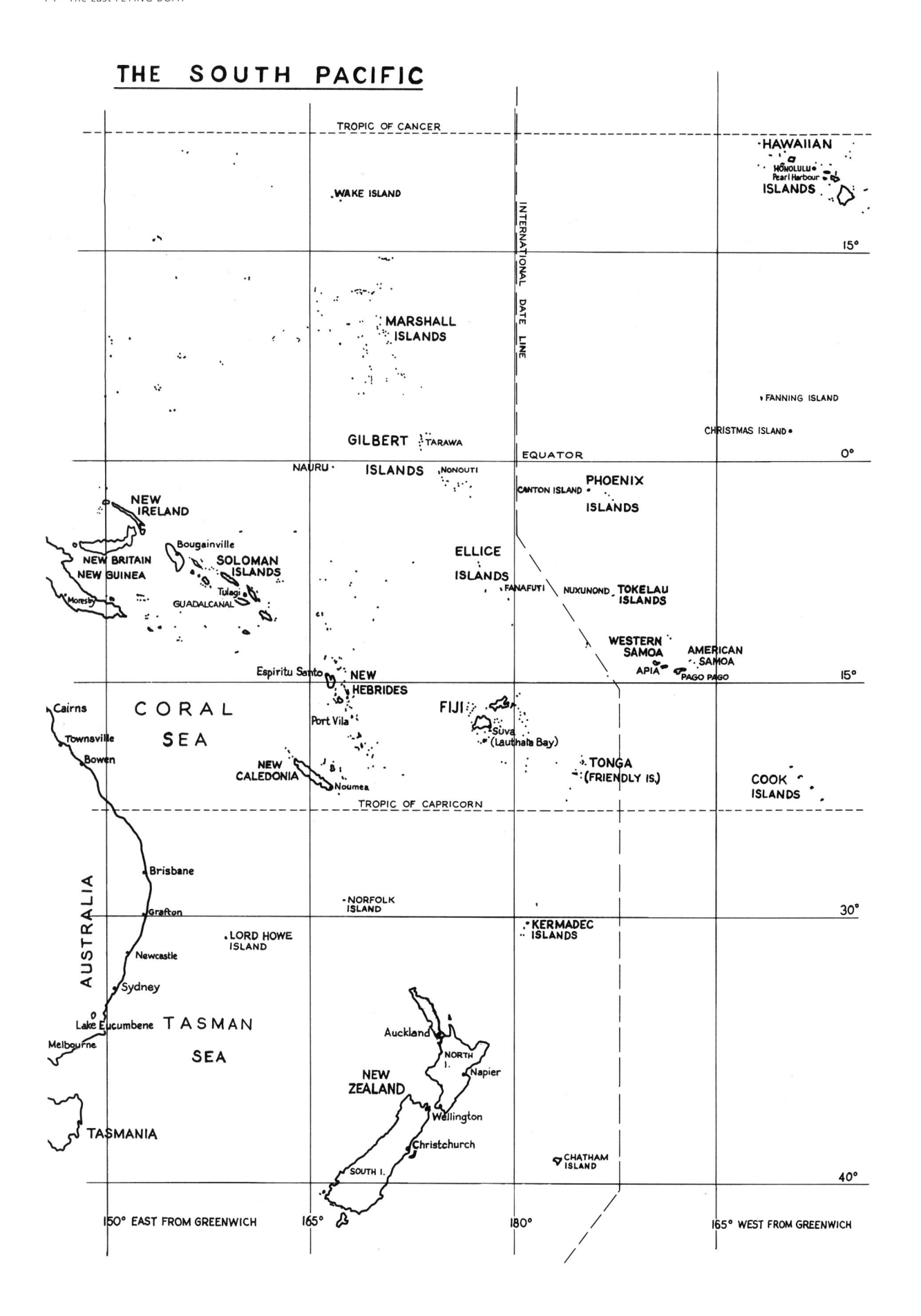

A Sunderland flying over the main entrance to the RNZAF Station at Lauthala Bay.

(Photo RNZAF official, via RNZAF Museum)

Total refurbishment it certainly was. It took a year, and when it was completed she was as good as a new aircraft. The Sunderlands were completely stripped, and any item not perfect was changed. Many skins were replaced if there was any hint of corrosion, particularly the planing bottom plating.

Not only did the RNZAF want their Sunderlands like new, they wanted them updated. Out came the old wartime radio equipment, in went new Marconi-built sets. This included two HF radio sets and two ADFs. Thus the New Zealand Sunderlands were easily distinguishable externally from other Mk Vs, by twin housings for the ADF scanners behind the astro dome, and twin aerial masts behind these.

All of the sono-buoy equipment was removed, but the Mark 6C radar was retained. As these Sunderlands were looked on as new aircraft, they were allocated new serial numbers, from NZ4105 to NZ4120 (the wartime NZ aircraft had carried the first four numbers in the series). So ML814 was no longer, she took on the new entity of NZ4108.

Two New Zealand crews came over to Calshot for conversion to Sunderlands, and two of the not yet completed Sunderlands, NZ4106 and NZ4107, were temporarily released for their training.

Thus, the first two aircraft to be completed and ready for delivery, by these two New Zealand crews, were NZ4105 and NZ4108.

May 21, 1953, saw NZ4108 delivered from Belfast to the Flying Boat Storage Unit at Wig Bay. There Fred Weaver, one of the two pilots of the Flying Boat Test Flight, took her up. On checking with his log book recently, he recalled the flight: "On the 26th May 1953 I flew NZ4108 on its initial test flight, prior to it being handed over to the RNZAF. I took along the NZ crew who were going to fly it home. It was a full test including Stalling, Critical Speeds, Asymmetric Flight, Steep Turns, etc., as well as checking all the equipment. Although I did not record it, (I did not always do so) I believe I finished with a single-engined flypast along the waterfront at Wig Bay — I do not remember the test flight encountering any problems with this aircraft."

Two days later Flight Lieutenant Tompkins signed the pile of large A3 sheets which made up the checking lists, accepting the aircraft and its equipment on behalf of the RNZAF (right down to the last cup - plastic, plate - dinner, and spoon - small). At last he and his crew were ready to set off on the flight which would take them more than halfway around the world.

The flight took 16 days to complete, although this included a three-day stopover at Seletar (Singapore), where the boat was given a check-over at the RAF base. The log book for May/June 1953 reads as follows:

29	May	Calshot — Malta	8 hrs 25 mins
30	-	Malta — Fanara (Canal Zone)	7 hrs 45 mins
31	-	Fanara — Bahrein	8 hrs 30 mins
1	June	Bahrein — Korangi Creek (Karachi)	6 hrs 10 mins

NZ4108 in the pontoon at Lauthala Bay, Fiji. Taken on 14 June 1953, immediately following her arrival from the UK. Standing on the pontoon on the far left is Flt Lt Vickers (RAF) and third from the left is Flt Lt Benseman (RNZAF).

(Photo RNZAF official, via RNZAF Museum)

2	June	Korangi Creek — China Bay (Ceylon)	10 hrs 00 mins
4	-	China Bay — Seletar	10 hrs 15 mins
9	-	Seletar — Soerabaja (Java)	6 hrs 00 mins
10	-	Soerabaja — Darwin	8 hrs 25 mins
11	-	Darwin — Cairns	6 hrs 30 mins
12	-	Cairns — Noumea (New Caledonia)	8 hrs 50 mins
13	-	Noumea — Lauthala Bay	5 hrs 15 mins

With a night stop at each of these ports, and eight or more hours of flying each day between them, the workload for the crew was tremendous. There were no special maintenance crew members aboard, just two pilots, two flight engineers, a navigator and a signaller.

Many of the en route stops no longer had regular calls by flying boats, which often made fuelling a real difficulty. The captain would have the problems of obtaining clearances for the oncoming day's flight, the flight engineers would have to carry out the daily inspections; sometimes there would be items to repair or replace, and all hands joined in to assist. All of this, combined with strange food along the route, together with political hot spots such as Indonesia to pass through, ensured a ferry flight that the crew would not quickly forget.

Despite all this the trip was relatively uneventful, apart from needing three attempts before they could get airborne at Soerabaja, due to a glassy water situation. On reaching Cairns the crew enjoyed the local king size prawns, and a good supply of fresh milk. New Zealanders are raised on dairy products, and doing without fresh milk on their flight rations had seemed a major hardship.

It was with an enormous sense of relief that Ted Tompkins and his crew reached 'home' — Lauthala Bay, near Suva, Fiji. The two new Sunderlands were very welcome arrivals at No. 5 Squadron, RNZAF, as the unit was down to its last two Catalinas. This would now be changing to six new Sunderlands!

A curious group quickly gathered around NZ4108 as she sat in the Braby pontoon. Barely had they arrived, when the ferry crew were given the task of flying the Governor of Fiji on a visit to the leper colony at Mokangai Island, about thirty minutes flying time away. It looked as if NZ4108 was in for a busy time in the Pacific.

A visit to Mokangai, however, was something a Sunderland crew would never turn down. At that time it was one of the last leper colonies in the world, and the personnel at Lauthala Bay had taken a special interest in it. They regularly raised money to assist the mission there, and a Sunderland visited every Christmas to deliver gifts.

Lauthala Bay had always been home to 5 Squadron. It had been formed there early in the war, equipped only with four ancient Short Singapore biplane flying boats — RAF 'throwouts' which their crews had great difficulty in coaxing from Singapore to Fiji. Later came Catalinas and, in December 1944, the four Sunderlands mentioned earlier.

Although No. 5 was primarily an anti-submarine and reconnaissance squadron, in post-war years these duties only accounted for a relatively minor part of the squadron's employment. New Zealand was responsible for the security of a huge chunk of the Pacific Ocean, and for law and order in the scattered island territories to be found within it.

From its Fiji base, the squadron was responsible for the surveillance of over 7 million square miles of ocean — by far the biggest 'beat' of any single squadron in the world. It stretched from beyond the Equator in the north to as far south as there was human habitation, or any suspicious maritime activity. From the French territories of New Caledonia and the New Hebrides in the west, it stretched eastwards to the Cook Islands and Tahiti.

In a way, the New Zealand Air Force was returning the Sunderland to the duties which had been in mind for it when it was on the drawing board, back in 1935. One envisaged role for the new craft had been that of a flying gunboat, to patrol the far flung reaches of Empire. On going into service in the Far East in 1938, the new flying boats had been fulfilling this role for barely a year when they were wrenched away by the demands of the Second World War.

Not that there were other colonial powers with hostile intent in the Pacific of 1953, or even hostile natives for that matter. And so the duties of No. 5 Squadron RNZAF were largely those of flying routine patrols and 'showing a presence' throughout its domain.

A young pilot with No. 5 Squadron at the time was John Laing, and he was to get to know NZ4108 quite well over the coming years. "I was second pilot to Flying Officer Jeff Patterson, and NZ4108 was allocated to Jeff as our aircraft. The main crew members were Jeff as captain, myself as co-pilot, Fg Off N E Richardson navigator, Sgt R Bleakley engineer, Sgt N Crump signaller, Sgt Galloway, armourer. I cannot recall the remaining crew members. We first flew NZ4108 as a crew on August 10th 1953, on an air test for 1hr 10 mins. Throughout the month of August 1953 we were under training so that we could convert from the Catalinas. The first flight of distinction flown by NZ4108 was to carry the Chief of Air Staff, RNZAF, Air Vice Marshall Carnegie, to Tarawa. She was commanded by Flight Lieutenant F J Vickers, an RAF officer on exchange for the purpose of converting us to Sunderlands."

Tarawa, some nine hours flying from Lauthala Bay, was a frequent call for 5 Squadron flying boats. It was capital of the Gilbert and Ellice Islands colony, almost on the Equator, and close to its intersection with the International Date Line. Tarawa is very typical of islands in this part of the Pacific. It is an atoll, comprising a series of islets on a reef 40 miles long.

Arriving by Sunderland, a visitor had a perfect view of the long crescent-shaped line of islands and the outlying reef, which together enclosed the lagoon. Once down on the lagoon — a vast inland sea — it became difficult to picture the geography of the place. No land is more than a few feet above sea level, and there is little more than a row of palms to be seen on the skyline.

From time to time, tiger sharks would gather in the lagoon, and the Gilbertese would catch them from small one-man canoes. A rope line and hook would be fixed to the canoe, and once a shark was hooked and had tired itself somewhat through frenzied swimming, the fisherman would club it to death. To a European flying into the Gilbert and Ellices for the first time, this was another world.

The Gilbert Islands had been occupied by the Japanese in World War Two, and Tarawa was the site of a particularly bitter five day battle in November 1943, when American forces pushed the occupiers out. After the war, the RNZAF had established a small station here, mainly to provide messing facilities for visiting flying boat crews. Normally a boat coming to Tarawa from Fiji would stay about five nights, patrolling the region and visiting outlying territories such as Nauru and Ocean Island.

After two nights at Tarawa, John Vickers and his crew left in NZ4108 for the return flight. This took over eleven hours, as they called at Funafuti en route. Funafuti, capital of the Ellices, is another atoll, and boasts the largest lagoon in the South Pacific. New Zealand maintained a met. station there, and so in support of this there were regular calls by No. 5 Squadron.

On return to Lauthala Bay, it was back to training flights for NZ4108 and her crew. John Vickers of course had a major role to play here, and he had an enormous depth of experience on Sunderlands. He had in fact been flying in 201 Squadron back in 1944, when ML814 was on strength with them, although he had not flown her then. Later, he had been a pilot on the North Greenland expedition of 1952. In Fiji, he became an easily identifiable personality through his RAF beret — he was the only person in Fiji to wear a beret.

John Laing continues his recollections: "On October 11th 1953, under the command of Flt Lt Vickers, 4108 flew to Canton Island and then carried out a search and rescue flight to Fanning Island, with myself as co-pilot. Canton Island, south west of Hawaii, was originally a staging post for the (then) Pan American Clipper flying boats on the route from Hawaii down to New Zealand. Fanning Island was an amplification station for the undersea telephone cable operated by Cable and Wireless between America, Fiji, New Zealand, and Australia. Wing Commander Le Pine, the CO at Lauthala Bay, was a passenger on this flight."

Canton Island has gone down in the annals of civil aviation history as the site of the Battle for Canton Island, which took place there just prior to the Second World War. It was perhaps the only territorial dispute ever to be centred around flying boats.

It was the era of trail-blazing for the trans-oceanic air routes. Juan Trippe of Pan Am had his mind set on opening a trans-Pacific route, but Britain's Imperial Airways had similar ideas. Small, unwanted, mid-Pacific islets suddenly became sought after staging posts. In 1937, the Americans had built a small unmanned lighthouse on Canton Island, as a way of staking their claim. In response, Britain claimed that the island had always been crown territory.

The 'battle' took place in 1939, when the American cargo vessel *North Haven* landed a construction party. Britain immediately sent a Royal Australian Navy cruiser to evict the intruders. After a period of stalemate, with feelings running high, a joint British\US control of Canton Island was established. Britain appointed a lone postmaster to take up residence on the island — their sole representative.[1]

During October, John Laing received his first dual instruction from John Vickers in 4108. Then, on 4 February 1954, aged 23 and a Pilot Officer, he was granted his captaincy and allocated NZ4110. Although 4108 was no longer 'his' Sunderland, John continued to fly her from time to time. From his log book he noted another typical 5 Squadron duty, in a flight he made on 1 July 1955 in 4108: "Flight to Yasawas, an island group in the north west of Fiji to uplift a Fijian woman with childbirth problems."

Just in these few flights, John has given a good picture of the type of work 4108 carried out during its two and a half years based at Lauthala Bay. Search and rescue work, administrative work (both for the air force and for the colonial administration), and medical evacuation flights.

Medical evacuation flights were regular occurrences for the New Zealand Sunderlands. In the South Pacific of the 1950s medical care was not the norm, but when childbirth complications arose, or serious accidents occurred, the value of

Carrying out engine runs on the slipway at Lauthala Bay. Note the fitter on the maintenance door by No 3 engine, carrying out engine adjustments.

(Photo RNZAF official, via RNZAF Museum)

hospital treatment was recognised. The only possible way of getting a remote islander to hospital, in a reasonable period of time, was by Sunderland. Even today, in the 1990s, many a Pacific islander no longer has the emergency medical services available to him that he did then, in the era of flying boats.

In one year alone, Fiji-based Sunderlands flew 40 mercy missions, and this engendered great goodwill towards the boats and their crews amongst the communities of the South Pacific. The crews, however, had some doubts about the reputation they were earning amongst their RNZAF colleagues. When asked their squadron, they were inclined to reply somewhat wryly, "Number 5 Maternity Squadron".

In between all this flying about, the occasional military exercise was squeezed in. Sometimes this was just a local anti-submarine exercise, sometimes Sunderlands were detached to exercise with the RAF boats at Hong Kong or Singapore, although 4108 never made such a trip.

Sometimes there were exercises with other Pacific powers and, in October 1955, the squadron exercised with the *Jeanne d'Arc*, the same French cruiser which NZ4108 had located when flying with 201 Squadron in 1944.

One such exercise in which John Laing flew 4108 took place during March 1955. He made the eight hour flight from Lauthala Bay down to Auckland, and then was flying her in the exercise for the next six days and nights. It was a joint anti-submarine exercise in conjunction with the Australian and New Zealand navies.

Another flight with NZ4108 vividly remembered by John Laing took place in January 1955. Fanning Island, where this episode took place, is one of three atolls making up the Line Islands. It was renowned for the dangerous coral heads in its lagoon. Another island of the group is Christmas Island, to become well-known several years later for the series of British and American nuclear tests held there...

"With fellow captain Flt Lt Torgersen, we flew to Canton Island and then off to Fanning Island for two nights. There are no moorings in the lagoon so we had to

Joint exercises. An RNZAF Sunderland in company with a Martin Marlin of the US Navy (VP 40), and a Lockheed Neptune of the RAAF (No. 10 Squadron).

(Photo RNZAF official, via RNZAF Museum)

anchor. What a night! We anchored in the entrance to the lagoon, and by 10 pm had three anchors out. NZ4108 rode most awkwardly, right up to the main anchor so that the chain was grinding against her hull. Later in the evening all the lights on shore went out so that, in the pitch black of a tropical night, we had no means of checking whether or not we were dragging anchor. No sleep that night.

"Bleary eyed, I greeted my fellow captain who had spent the night ashore. 'Oh,' he said, 'John, you worry too much. Take your anchor watch ashore and I'll see you in the morning'. Next morning, January 19, out to 4108 and there was Torgersen, no sleep for him either. 'Lets get out of here, you were right, John. This is a shocking place.' So it was back to Lauthala Bay in 4108."

Other regular ports of call for 4108 were the Tokelau Islands and Western Samoa. The Tokelaus are actually part of New Zealand, while in the fifties New Zealand administered Western Samoa under United Nations mandate. Thus frequent administrative flights were needed to both territories.

Western Samoa, popularised and romanticised by Robert Louis Stevenson's period of residence there, was a favourite tasking for the Sunderland crews. Satapuala Bay was the mooring here, and for the New Zealanders that meant at least one night at Aggie Grey's. In the South Pacific, Aggie Grey's was everything and more that Raffles was to Singapore. Not only was there the legendary watering hole, but at Aggie Grey's there was also a legendary lady to go with it.

In those days, Aggie's was still a hotel in the traditional South Seas style — a white painted timber building surrounded by shady trees on the waterfront at Apia. A cool and inviting oasis to the thirsty Sunderland crews, or any other seafarers; its open corridors were air-conditioned only by the trade winds.

A good idea of NZ4108's routine work while based at Lauthala Bay can be gained from No. 5 Squadron's diary and, as a sample, here are the entries relating to 4108

John Laing leaning from the captain's window of NZ4110 at Labasa, during the Joyita search.

(Photo W K Cowan)

for one month only (June 1955). This is not in any way a complete listing of the flights she made during that month, but merely notes on some of the highlights. The details preceeding each entry are the date of the flight, the captain, and the duration. [2]

3-6-55 — Flt Lt Noble-Campbell. 5.50 hrs. Carried the harbourmaster Capt Harness on a survey of the buoyage system in the Koro Sea, followed by a supply drop and short range navex to Kabara Island.
7-6-55 — Flt Lt Noble-Campbell. 5.15 hrs. Carried out a search in the Koro Sea area for a suspected poaching fishing vessel, part of a fleet carried by the Japanese ship Tenyo Maru. The chaser was not found.
14-6-55 — Flt Lt Noble-Campbell. 8.20 hrs. A flight to Tonga with Ministry of Works personnel, returning to base via Vavau. On the return flight the active volcano of Fanua Lai was investigated and photographed.
16-6-55 — Flt Lt Noble-Campbell. 6.00 hrs. A further flight to Tonga to uplift the Ministry of Works officials.
20-6-55 — Fg Off Laing. A research flight carrying Mr Hall, an astronomer, to view an eclipse of the sun.
23-6-55 — Flt Lt Noble-Campbell. 9.25 hrs. A night navex. The RMS Himalaya was intercepted at midnight as planned. Mr Blackwell, scientist, was also on this aircraft to take the fourth and final photograph of the Zodiacal Lights.

The end of 4108's deployment at Lauthala Bay was not now far off but, before then, she was to take part in what became one of the most notable search operations ever carried out in the South Pacific. On 3 October 1955, the motor vessel *Joyita* left Apia, Western Samoa, for the Tokelaus, some 270 miles to the north. With 25 people on board, she failed to arrive at her destination, and was declared overdue on the 6th.

On the 7th, a search was mounted, with NZ4109 being despatched from Lauthala Bay. The next day, 4108 under Fg Off McGrath joined her, and spent three full days on the search. Flt Lt Torgersen with 4110 was also involved, the boats returning each evening to Satapuala.

The Joyita. The mystery of its disappearance has never been solved.

(Photo W K Cowan)

After one day back at base, Fg Off McGrath and 4108 returned for another three full days of searching, with still no trace of the missing vessel. By the 13th, when 4108 finished its involvement, the search had been going on for a week, with three New Zealand Sunderlands involved throughout that time. The search was called off on the 14th. There were smaller searches by individual Sunderlands over the weeks to come in more distant areas, usually following-up a reported possible sighting, but without result.

Then, on 10 November, the master of another inter-island ship sighted a waterlogged vessel 450 miles WSW of Apia. Investigation proved it to be the *Joyita*, listing 55 degrees, and without a trace of the 25 people who had sailed on her. Thirty-six days had passed since her initial disappearance. A further Sunderland search was mounted for survivors, but proved fruitless and was abandoned on the 16th.

It was realised that during the original search, several Sunderlands must have passed within radar range of the *Joyita*. Tests were carried out on her, and it was discovered that she possessed natural stealth characteristics. She could only just be detected on radar at 1 to $1^{1}/_{2}$ miles, whereas most vessels of her size would be detected at 20 miles.

On 25 November, Sqn Ldr Gunton flew 4108 to Wallis Island to collect some wreckage found there. Yet no trace was ever found of the 25 people who had sailed in the *Joyita*, and their disappearance remains one of the great mysteries of the South Seas.

January 5, 1956, saw 4108 fly out of Lauthala Bay for the last time — she was having a change of squadron. No more picking her way between the coral heads of distant lagoons. No more met. flights over the open and threatening Pacific, to track down newly-forming cyclones. But there would still be remote islands, search and rescue flights, even mercy missions. And of course, still the never ending round of training flights. She had been transferred to No. 6 Squadron, Hobsonville. The change of squadron brought the inevitable change of identification letters, KN-B being replaced by XX-D.

No. 6 was the 'home' flying boat squadron of the RNZAF, its Hobsonville base being on the Waitemata Harbour, Auckland. The Squadron was a mixed regular and territorial unit, although during 1957 it was replaced by the Maritime Operational Conversion Unit.

Obviously in a squadron of this nature, training flights made up a large part of the flying, but — not surprisingly for a Sunderland unit — there were interesting variations. In June 1956, John Laing flew 4108 on a search and rescue mission out of Napier, looking for a missing fishing boat. (John had been posted to the Maritime Conversion Unit.)

A frequent task for Hobsonville Sunderlands was a flight to the Chatham Islands. Chatham Island and nearby Pitt Island are 470 nautical miles east of New Zealand's South Island. Both support small farming communities, and pre-war, the only transport to the islands was by the small coaster *Holmewood*. In November 1940, the war suddenly came much closer to New Zealand when the *Holmewood*, having just left Chatham Island, was sunk by a German commerce raider.

This episode shocked the government into establishing flying boat access to the island, and locals surveyed and buoyed (with oil drums) two runways on the island's shallow Te Whanga Lagoon. After the war, Tasman Empire Airways Limited (TEAL — now Air New Zealand) opened a service with its Solent flying boats. There were normally five flights a year, mainly in the summer, and primarily to provide transport for the island's children to mainland boarding schools.

TEAL's Solents discontinued the service in 1954, and the air force Sunderlands then made occasional flights to the Chathams as needed. On 23 August 1956, John Laing took 4108 to Chatham Island, to bring out a woman with childbirth problems to hospital in Auckland. It was back to maternity flights for the old boat!

To update the Chatham Island story, NZ4111 hit a rock in the lagoon in 1959, was beached, stripped, and its hull is still on the island. A land airport was constructed there in 1981. In 1990 the coaster *Holmedale* made her last voyage to the Chathams, due to the NZ government discontinuing its subsidy.

During 1957, the RNZAF decided that it had more Sunderlands in service than its current needs required, and the decision was made to place several of them in long term storage. One of those selected was 4108 (probably because of concern over corrosion of her port wing spar), and she was placed in long term storage at Hobsonville on 21 June 1957.

Maintenance staff had been aware of the corrosion, on the lower front spar boom, for some time. Measurements had been made, and Shorts at Belfast consulted on the problem. The boat had been cleared to continue flying, but a daily inspection had to be made of the area.

Once again, it looked very much like the end of the old boat's flying days. But she was destined for yet another reprieve, as the following chapter will reveal. And of the other New Zealand Sunderlands? NZ4117 suffered a similar fate to 4111. She was holed on landing at Tarawa in 1961, and later broken up. John Laing made the last flight in a New Zealand Sunderland on 2 April 1967, when he flew 4107 from Lauthala Bay back to Hobsonville (it was also the last flight by a military Sunderland anywhere in the world).

With the exception of 4115, the remainder were sold for scrap and broken up. In 1993, NZ4115 was in the care of the Museum of Transport and Technology, Auckland — although sitting outside and in a very sad state of repair.

Footnotes Chapter 5.

[1] *Seawings* — Edward Jablonski (p 109).
[2] National Archives of New Zealand — Air 143/5.

Chapter six

Flying Boat Airlines

The next chapter in ML814's story takes us from New Zealand, across the Tasman Sea, to Australia. Before looking at her arrival there, we will review the role which civil flying boats played in Australian commercial aviation in the years following World War Two. It was, in fact, a far more important role than they played in most other parts of the world.

The possibilities of the large flying boat as a reliable means of long distance transport was first demonstrated to the Australian public in 1928, when four Supermarine Southamptons of the RAF's Far East Flight flew there from the UK. After a flag-waving circumnavigation of the continent, making leisurely calls at all the major cities, they set off for the long homeward flight.[1]

The boats and their crews had been warmly received everywhere by an air-minded Australian public, although the somewhat notorious Sydney press gave them a hard time after deciding that they were "stuff-shirted pommies". The crews left Australia, however, remembering as typical the comment of their civic host at Port Headland: "You was bloody good blokes, even if you was bloody officers."

The flight had made a deep impression on many people as to the future possibilities of air transport. The four large boats had flown from the UK to Australia, around the continent, and back, without any major mishap or breakdown — each having been a self-contained, self-supporting unit.

Flying boats really became a part of the Australian aviation scene in 1938, when Qantas commenced joint operation with Imperial Airways of the Empire Boat service from Southampton. Of the thirty new Empire Boats ordered from Shorts at Rochester, six were to be owned by Qantas. Construction of the new flying boat base at Rose Bay on Sydney Harbour was started. Qantas even bought a 12-foot sailing dinghy for pilot training. [2]

There were to be three flights a week, each boat operating the full length of the route, so that passengers did not have to change. Qantas crews flew from Sydney to Singapore, Imperial Airways crews taking the remainder of the route. The Empire Boat *Cooee* left Rose Bay on 5 July 1938, to commence the first flying boat through-service to Southampton. Australia was now joined to the UK by a regular, reliable and comfortable air service.

The second war brought about a big increase in the number of Australians conversant with the operation of flying boats. In July 1939, the first group of RAAF aircrew embarked for the UK, to commence conversion training on the new Short Sunderland flying boats on order for the RAAF. These boats were to be operated by the newly formed 10 Squadron, RAAF. While the crews were still working up at Pembroke Dock, the European war broke out and the Australian Government put the Sunderlands and their crews at the disposal of the RAF.[3]

During the war, several more Australian flying boat squadrons were formed, and Number 10 went on to become the most notable Sunderland squadron of the war. Although 10 Squadron remained based in the UK until the end of the European war, many of its personnel transferred home, to help build up several squadrons of

Catalinas and Sunderlands which played an important role in the Pacific war. When the war ended, as in other countries, there were a number of Australians keen to turn their aviation experiences into peacetime enterprise. Due to the circumstances of suitable and available aircraft, and routes, much of this post-war enterprise in Australia involved flying boats.

From this period, three names stand out — P G (Bill) Taylor, Stewart Middlemiss, and Bryan Monkton — each instrumental in their own way in the establishment of commercial flying boat services. P G Taylor's flying career started during the First World War. He applied to join the Australian Flying Corps, but there were no vacancies. Undaunted, he went to England where he was readily accepted by the RFC. At the time, he was unaware that this was due to their high loss rates in France.

The war over, Taylor obtained employment as a commercial pilot, eventually becoming a captain with Australian National Airways, the airline set up by two trail-blazing Australian pilots, Charles Kingsford-Smith and Charles Ulm. In addition to airline flying, P G Taylor flew as navigator with Kingsford-Smith on various pioneering flights, including the first crossing of the Tasman Sea. Taylor had taught himself astro-navigation, and became a pioneer in air navigation.

During the second war, Taylor was involved in survey work, much of it in Catalina flying boats. This was the beginning of what developed into PG's great love of flying boats. In 1951, he used his influential contacts in high places to get the use of an RAAF Catalina, and in it made the last great air crossing of an unconquered ocean — from Australia, via Tahiti and Easter Island, across the South Pacific to Chile.

With the successful South Pacific crossing behind him, P G (now Sir Gordon) Taylor realised that the pioneering days were gone. He looked for new ways of combining his love of flying boats with his knowledge of, and feeling for, the peoples of the South Seas islands.

In the post-war prosperity, he could see a growing demand for tourism to the South Seas. But regular services at that time could only take tourists to a few commercial centres on the edge of European life. Such visitors, he knew, could never really taste island life. He evolved the idea of a touring flying boat, sailing freely to the remotest islands, alighting, and waiting at anchor in the safety of the lagoon while its passengers took in the pleasures of island life.

To bring this dream to reality, Sir Gordon Taylor flew to England in 1954. There, at Hamworthy, near Poole, he found a collection of Sandringham flying boats lying derelict on the slipway. They had been made redundant by BOAC's move out of flying boats, yet some of them were almost new.[4]

The sight of these magnificent craft, lying unwanted, really drove home to Taylor the fact that the days of the flying boat on the international airline routes had passed. This thought did not, however, deter him from purchasing one of the Sandringhams, as he knew it would find a role in his niche operation. It was named *Frigate Bird III*, and taken to Saunders-Roe at Cowes to be prepared for the homeward flight.

Operating out of Sydney's Rose Bay Water Airport, *Frigate Bird III* carried many passengers on unforgettable South Seas cruises over the next four years. It was a remarkable operation, a 'one man airline', operating without any expensive back-up or support facilities. Taylor employed only a first officer, radio operator, flight engineer, steward, and stewardess — the essential crew. The flight engineer carried out necessary maintenance between cruises.

Taylor's cruises covered the vastness of the South Pacific, calling at remote islands and atolls, where often his passengers were the first European tourists ever to call. He had many contacts in the islands, both expatriate and native, and he could use these friends to welcome and play host to his tourists.

P G Taylor's public school background, his insistence on perfection, and a rather egotistical manner was at times trying on his crew, thus leading to a number of amusing tales relating to him.

On one occasion Dick Froggatt, his bluntly-spoken flight engineer, was changing a starter on the lagoon at Bora Bora with the assistance of Harry Purvis, the first officer. They were interrupted by Johnny Green, the steward, coming alongside in the dinghy with a message. "The captain says he will be coming aboard with several VIPs at about two o'clock, and will expect everything to be in order."

An exasperated Dick, covered in sweat and oil, stuck his head out from the nacelle. "Oh, he will, will he? And will he be needing the dinghy, or will he be walking across the bleeding water?"

After a number of years of successful operation of his flying boat cruises, various factors made Taylor consider winding up his venture. Then came an offer to purchase *Frigate Bird III* by the French airline TAI, to operate the last leg of their planned new route from Paris through the Orient and Noumea to Tahiti.

Their senior South Pacific pilot, Captain Allais, had recommended the aircraft to his company, and Taylor agreed to sell. However, before he could take charge of his new flying boat, Captain Allais was killed in a Catalina crash while attempting a glassy water landing. Nevertheless, the deal went through in time for TAI to use the boat (which they always called *Bermuda*) to open the new service in October 1958.

Sir Gordon Taylor was a rugged individualist who did not fit well when working as a captain for other airlines. But he was a great pioneer, an innovator and perfectionist in the field of navigation, which to him assumed the proportions of a religion. An outstanding proponent of the values of the flying boat, he was often fêted more in various other lands of the Pacific than in his home country, Australia.

The second 'flying boat entrepreneur' in post-war Australia was Captain Bryan Monkton. In 1946 he bought four Sunderlands and assorted spares from the RAAF flying boat base at Rathmines in NSW. With his purchase, Bryan Monkton set up a new airline known as Trans Oceanic Airways, operating from Rose Bay.

P G Taylor became a director of the company and assisted with its establishment, as well as flying as captain. Trans Oceanic operated to Lord Howe Island, New Caledonia, New Hebrides, the Solomon Islands, New Guinea, Hobart and Grafton. Trans Oceanic's main competitor was Qantas, which was still heavily involved with flying boats in the early post-war years. At first, there were Hythe class boats on the London route, but these were quickly phased out. Catalinas were used extensively on the Pacific Island routes, to which were added five Sandringhams in '50 and '51. The boats were by now only used on the South Pacific and New Guinea routes. [5]

In 1947, Trans Oceanic Airways inaugurated services to Lord Howe Island, and a little later were followed by Qantas. This meant that the two airlines were flying in parallel on routes to Noumea, the New Hebrides, Solomon Islands and Lord Howe Island. Although there were always friendly relationships between flight crews, the rivalry was more bitter in the higher echelons of the companies.

Left and below: Passengers coming ashore from a Trans Oceanic Airways' Sunderland at Lord Howe Island, in the late forties.

(Photo Lord Howe Island Museum)

An incident demonstrating this feeling occurred in June 1949. Both airlines had flights due into Lord Howe Island, but they were delayed by bad weather. Then the forecast predicted a quieter spell, allowing a flight in on the evening tide and out with the tide next morning. The Qantas Catalina arrived first, and took the shallower northernmost mooring, leaving the deeper mooring for the bigger Sunderland. Half-an-hour later the Trans Oceanic boat arrived, touching down on the lagoon just on sunset. After seeing his passengers ashore, Captain Phillip Mathiesen and several of his crew elected to spend the night aboard, in view of the strong westerly being forecast.

Holiday-makers boarding their Sunderland for the return flight to Sydney, on the lagoon at Lord Howe Island.

(Photo Lord Howe Island Museum)

Although the crew had an epic night on the Sunderland, running engines to relieve the strain on the moorings and rigging new mooring lines on various occasions throughout the night, at least the Sunderland was still in one piece by the morning. Phil Mathiesen then made a hasty departure from the Island: "We got our passengers aboard, but the wind was so strong that I was unable to turn downwind. I let the aircraft sail back until the stern was almost on the beach then took her off, straight across the lagoon, and on to Sydney." [6]

Qantas, however, had not fared so well with their unmanned Catalina. It was up on the beach and a write-off. Phil Mathiesen continues: "Well, there were some snide remarks by various people that I had rowed across and cut it loose that night, but I can assure you that I had more on my mind than rowing across to the Catalina."

Then, just two months later, came an event described by Hudson Fysh, chairman of Qantas, as "One of the most sensational happenings in our long history". A Qantas Catalina had returned from the Lord Howe run, and was lying overnight at its Rose Bay mooring. There was an explosion, the boat burst into flames, then sank quickly. As no one had been injured, the immediate reaction was one of slight amusement, until a few weeks later Bryan Monkton was charged with the crime.[7]

The police produced in court a piece of equipment consisting of a cheap alarm clock (which they alleged Monkton had purchased at a local shop), a battery and an induction coil, all of which were bound together on a piece of wood by ordinary household string. They alleged that this device had caused the explosion; but it was pointed out by an explosives expert that all the parts were in good condition, the glass on the clock unbroken, the string and wood unsinged, and that they could not possibly have been at the centre of an explosion and subsequent fire.

Monkton stated that the smaller Qantas Catalinas were no real competition to his larger Sunderlands, and that neither he nor anyone else in TOA had any real motive to try to "sink the opposition". The twelve man jury considered their verdict for only a short time and, in Bryan's words, "Justice prevailed and I was acquitted".

The whole episode left a nasty taste in the mouths of the independent operators. They felt it was all part of a government policy to force out the small companies, as they were making life difficult for the nationalised concerns. This feeling of injustice was heightened by the fact that the smaller independent companies tended to be owned and staffed by ex-servicemen, who felt that they had 'done their bit' for the country in the recent war, and therefore deserved some consideration.

Despite this incident Bryan Monkton continued with TOA, although financial changes later removed him from control. Gradually life became more difficult financially for the airline, until in 1953 it went into liquidation.

The third post-war independent flying boat airline in Australia was Barrier Reef Airways (BRA). This company had been set up in 1946 by Stewart Middlemiss, a friend of Bryan Monkton. In partnership with Chris Poulson — a retired sea captain and proprietor of Heron Island on the Barrier Reef — Middlemiss hoped to cater for and encourage tourist traffic on the Barrier Reef. Two ex-RAAF Catalinas were bought for £500 each and converted for civil use.

Even in those days, the beginnings of Australia's 'two airline policy', which continued into the eighties, was beginning to be felt. This policy encouraged the state airline, Trans Australian Airways, and one competing private enterprise airline, on all internal routes — to the exclusion of all others. As, however, the flying boats did not compete with TAA's landplanes, operating rights were granted. In 1947 BRA commenced services to Heron, Lindeman, and Daydream Islands.

The newly established airline was struck a blow at the end of 1947, when Chris Poulson was lost at sea from a yacht. Stewart Middlemiss carried on alone, expanding his fleet by purchasing two Sandringhams from New Zealand (TEAL). To finance this deal, however, he was forced to sell a 51% share in the company to Ansett Airways, the small but expanding landplane airline.

Eventually Reginald Ansett, ever the astute businessman, put the pressure on so much that Stewart Middlemiss was forced to relinquish all his BRA holding in exchange for Ansett Transport Industries shares, plus a directorship of the company. Middlemiss, who had started his aviation career with Ansett in pre-war days, commented "Gradually I was getting to call Ansett 'boss' rather than Reg".

Barrier Reef Airways, now fully owned by Ansett and trading under the name of Ansett Flying Boat Services, took the opportunity to acquire the Trans Oceanic Airways assets when that company folded in 1953. These included rights at the Rose Bay flying boat base, enabling the Ansett operation to move from Brisbane to Sydney. Before long, Ansett was the last Australian flying boat operator, and their routes rapidly contracted. In 1953, they had been operating from Cairns southwards through Brisbane and Sydney to Hobart, as well as their South Pacific charters and the Lord Howe Island service. Gradually the services between the mainland centres were dropped.

Following Ansett's takeover of the much larger Australian National Airways in 1957, it became policy to discontinue all flying boat services which competed with landplanes. This only left the Lord Howe Island route and the occasional South Pacific charter. Life at the previously busy Rose Bay base on Sydney Harbour became much quieter, with just two Sandringhams retained to operate the essential service to Lord Howe Island. These two boats were VH-BRE, the ex-Qantas *Pacific Chieftain*, and VH-BRC *Beachcomber*.

This service continued without incident until July 1963, when VH-BRE was on a charter flight to Fiji. It was taking a team of Australian bowlers, and making stopovers at Lord Howe Island and Noumea en route. Stewart Middlemiss was captain: although running the company at the time, he still kept his hand in as a pilot. Whilst they were night-stopping at Lord Howe Island on the outward journey, a very strong wind blew up.

Captain Middlemiss was ashore in one of the lodges for the night, when he was awoken by a banging on his door at two in the morning. "Hey skipper! Your flying

Pacific Chieftain sinks at its mooring in the lagoon.

(Photo via Noel Hollé)

boat is up on the beach." He made his way down to the lagoon, and was greeted by a sad sight. He later described the situation in an interview. [8]

"A heavy sea was pounding in, and the flying boat had been washed up on the beach where it was resting on the wingtip, which had bent. It was an awful mess. My first officer was Ron Gillies, who had been a Wing Commander in the RAAF, flying Sunderlands in Britain during the war. He was quite famous and has been mentioned in a few books. I said to him that I thought we ought to leave the aircraft on the beach and sandbag it, until we could do something with it.

"'No' he said, 'More waves will come in and damage it further'. So with the aid of the passengers and the hotel staff, we dug a bit of a channel for it. The wing was badly damaged but the engines were all right and I started them up. It was now fairly calm and conditions were not bad, and we managed to put the Sandringham back on the mooring. The float was off the starboard side, so we packed the port wing with sandbags to stop it from swinging around. Blow me down if the gale force wind did not come up again that night! The Sandringham collapsed and sank in the lagoon. The salt water and battery acid ruined it."

Pacific Chieftain, it was decided, was a write-off. She was refloated and towed back to the shore to allow the stripping of salvageable items. Engines, cabin fittings, control surfaces, and many other items were stripped from the Sandringham, then several weeks later she was towed out of the lagoon into the deep Pacific and scuttled.

Islanders, sad at her loss, sang and threw wreaths and leis on the ocean as the old girl went down. As the flying boats provided the vital link with the mainland, they were looked on as belonging to the island. Now there was only one left. Ansett Flying

Pacific Chieftain on the beach and stripped of usable components, ready to be scuttled.

(Photo Lord Howe Island Museum)

Boat Services faced a considerable problem. They had to find a replacement, as one boat could not maintain the service alone. They were obliged to continue the service, and received a government subsidy to do so. Flying boats had by now virtually disappeared from passenger routes the world over, and there was just not a replacement to be found.

Stewart Middlemiss even tried RAI, the French-owned Pacific subsiduary of TAI, which had bought P G Taylor's *Frigate Bird III* back in 1958. Now manned by a French-speaking British crew, it was still in use and they would not part with it. (Interestingly, when it did finally finish service, the French carefully brought it back to France and put it on display in the Musée de l'Air at le Bourget.)

Then it was realised that the RNZAF had some Sunderlands in long term storage, and one of these might be suitable for conversion. Ansett approached the Department of Civil Aviation, which in turn approached the New Zealanders and approved a deal. The deal between Ansetts and the New Zealand government was for the purchase of NZ4108, including four engines, but without radios or any military equipment. The price was NZ£20,000, and Ansetts must pay for any overhaul work carried out by TEAL. These arrangements were not revealed at the time and, over the years since, many incorrect figures have been quoted for the price paid. Often these figures confused purchase price with the later cost of conversion work.

There was one rather curious condition attached to the purchase deal. In the event of any hostilities arising in which Australia and New Zealand were involved, the Sunderland would have to be returned to New Zealand. Remarkable — considering that by then Sunderlands were 25 years old as a military type, and if returned to New Zealand, this one would be in the form of a passenger aircraft.

Contacts with New Zealand were not new to Stewart Middlemiss. Over the years he had provided Sandringhams on charter to operate the Chatham Islands service when RNZAF Sunderlands had not been available. Ansett had even flown the Coral Route for TEAL when their Solents were unserviceable.

The end of a flying boat. Pacific Chieftain is scuttled off Lord Howe Island.

(Photo Lord Howe Island Museum)

A trip was made to Hobsonville to examine the Sunderlands in storage there. They sat outside, engines and radio equipment removed, but otherwise intact and unprotected from the elements. NZ4108 was selected, mainly on account of the low number of hours she had flown. Ansetts were told of the problem of wing spar corrosion, but they considered this to be acceptable. The boat was taken down the harbour to the TEAL base at Mechanics Bay, to be prepared for the flight to Australia, as well as having her military gear removed.

Although TEAL had finished their own flying boat operations in 1960, the big Mechanics Bay base had been kept on. It was still an administrative centre, work on components from their landplanes was done there, and they carried out all major maintenance work on the RNZAF Sunderlands.

On 18 December 1963, the Ansett crew arrived to ferry their 'new' boat home. It consisted of Captain Stewart Middlemiss, Captain Lloyd Maundrell (Chief Pilot of AFBS), Captain Hank Henry (Operations Manager — who was to act as bowhand), and Ron Bush (Chief Maintenance Engineer and Flight Engineer).

Lloyd Maundrell had flown with the RAF and RAAF during the war, and with BOAC and Qantas on the England — Australia route after it. In 1952 a friend had invited him to 'help out' on the flying boats at Rose Bay. He started working with Trans Oceanic Airways, liked flying boats so much that he stayed with them, and transferred to Ansetts when Trans Oceanic folded. By now he had worked up to being Chief Pilot.

The crew went aboard NZ4108, which was moored near the jetty at Mechanics Bay. Interested in seeing the big departure, the entire TEAL staff turned out on the jetty to see them off. Lloyd Maundrell takes up the story: "Stewie Middlemiss was pretty rusty, but he wanted to put on a show for them, and so he elected to fly it to Rose Bay. Hank Henry, all 20 stone of him, was in the bow, and he had the aircraft on short slip. Stewart started up the two outer engines, and gave Hank the clearance to slip the

Captains Stewart Middlemiss (left) and Lloyd Maundrell at the controls of NZ4108 for the ferry flight to Australia.

(Photo Aviation Heritage)

mooring, which he did. However, as Stewart opened up the throttles to taxi, there was an almighty thump, and the aircraft's bow went down. We were all flung forward, and Hank disappeared through the hatch, out into the water. The rest of us were picking ourselves up to try and get back into our seats. Ron Bush, the flight engineer, had forgotten to remove the storm pendant." (This is a heavy cable attached underwater from the bow of the flying boat to the mooring buoy.)

"If you could have seen the look on the TEAL spectators' faces, it was one I shall never forget... After eight hours and four minutes of uncomfortable flight we arrived at Rose Bay."

No doubt the embarrassment was felt all the more keenly because of colonial rivalry, and the desire of the Aussies to put on a show for the Kiwis. Added to which, as the crew was composed of the company's 'top brass', the event would take a bit of living down when news reached Australia.

Once NZ4108 was up the slipway at Rose Bay, the pressure was really on to complete the civil conversion. Five months had already elapsed since the loss of *Pacific Chieftain*, and everyone knew that *Beachcomber* could not be expected to continue the service reliably on its own. It was only a matter of time before the need for major work on her would ground the service.

The Sandringhams had all been built by Shorts at Belfast in the late forties, by converting military Sunderlands to civil airliner standard. Their internal layouts, although varying somewhat to suit customers' individual requirements, generally followed the style of the pre-war Empire Boats. This consisted of many (six or so) small cabins, with seats facing each other — reminiscent of a railway carriage compartment.

Ansetts opted for a more conventional airliner layout, with all forward facing seats. A full upper deck was constructed to the rear of the flight deck, with stairs leading up to it from near the rear entrance lobby. There were two cabins on the lower

Work continues in the hangar at Rose Bay. Fairing in of the front turret and enlarging the forward entrance hatch.

(Photo Charles Shiplee)

deck, with the galley amidships separating them. Thus the galley was still in its original position as on the Sunderland, with its opening hatches retained for use as emergency exits, and to allow better airflow through the boat when on the mooring. Similarly, the upper gun hatches were retained as emergency exits for the upper cabin.

As Rose Bay was an overhaul and maintenance base, there were no design engineers on hand to supervise the work. Thus Ansett's engineering section from Essendon Airport, Melbourne, took design responsibility. But the work itself was carried out by Rose Bay staff. The external changes were (and still are) fairly obvious. Fairing in of the turrets, removal of the bomb doors and racks, enlarging of the forward entrance door. The door changes were easy, as Ansetts had a Shorts 'mod' kit for a Sandringham. Less obvious, but more demanding, were the internal structural changes. The rear cabin floor had to be lowered, and the entire upper cabin floor constructed.

Although in general the Australian Department of Civil Aviation (DCA) accepted the work to be to the same standards as in the original Sandringhams, in some safety aspects they went beyond that. The seat structures had to meet the then current airliner standards of stressing to 7G. This was considerably stronger than in a Sandringham, involving the fabrication of heavy seat rails. The seats themselves were modified slightly from those fitted to Ansett's Fokker F27s, and were built by Ansett's own coachbuilding subsidiary, Ansair.

The job was made easier by utilising many components salvaged from *Pacific Chieftain*. For example cabin fittings, the Janitrol heating system and air ducting, the radios and flight instrumentation, and much else, were re-used.

While these modifications were being made, the opportunity was taken to completely refurbish the airframe, and all electric wiring was renewed. Changes were also required to the power plants. It is vital in a civil aircraft that these can be changed quickly. But in the Sunderland the oil coolers were remotely mounted in the wing leading edges. In the event of an engine failure there is usually oil contamination, necessitating changing or flushing of the coolers. Thus the oil coolers were moved and slung below the engines, giving a 'QEC'-type unit.

Inside the hangar at Rose Bay — an engine is being changed on Beachcomber.

(Photo Harry Williams)

Noel Hollé, Chief Inspector at Rose Bay throughout the time of Ansett Flying Boat Services, describes the work going on at the time. "DCA were quite co-operative, but of course it was in their interests as well to have two boats operational. BRC did a marvellous job carrying the service through the autumn, winter, and spring of '64 — it was trouble free. Of course the winter is the off season to Lord Howe, and the service in July and August got down to about once a fortnight. It kept the conversion work progressing at a non-stop rate.

"The workshop at Rose Bay was fully approved, the metal workers were able to fashion any of the fixtures, and the stringers to support the floor. Even the cross members were able to be fashioned at Rose Bay. All that was needed was the technical expertise of the engineering staff at Melbourne — they supplied the drawings which were immediately approved by DCA."

Even old flying boat stalwarts like Noel, who swore by the reliability of 'the boats', was amazed by the performance of VH-BRC (*Beachcomber*) throughout this period. Fourteen months had now passed since the loss of *Pacific Chieftain*, and the service had been maintained without a hitch. *Beachcomber* had not suffered a single cancellation through unserviceability.

September 1964 saw the conversion work nearing completion. Most observers at Rose Bay were of the opinion that the new boat wasn't as pretty as a genuine Sandringham. The big, bulbous nose closely followed the contours of the original

Engine maintenance in the hangar at Rose Bay.

(Photo Tony Hutchings)

turret, and did not look as pleasing as the Belfast factory's streamlined nose. But from Ansett's point of view, it was far more practical. There was lots of bulky freight on the Lord Howe Island route, and it was easy to load it into the big nose.

When it was put on the Australian register, as VH-BRF, they called it a Sandringham. Everyone knew that it really wasn't, but it made the paperwork easier. Besides, from a technical and operational viewpoint, it was identical to one.

Pristine and gleaming in the white and green livery of Airlines of New South Wales, she was christened *Islander* by Mr C A Kelly on 27 September 1964. An obvious name, the choice of Stewart Middlemiss, but a more suitable one it would have been impossible to find.

The next day, she was up on her first test flight under the command of Lloyd Maundrell. Just a short trip of an hour-and-a-quarter, but enough to show that there were no major problems. Then came two busier days of performance trials, with Captains Bill Wilcher and Ron Gillies acting as First Officers for this work. Surprisingly, she turned out to be slightly faster than *Beachcomber*. Possibly that ugly bulbous bow was more aerodynamic than the pretty one. Possibly it was just that after repairs from many scrapes on the coral at Lord Howe, *Beachcomber*'s hull was not as clean as it should have been, and was giving her extra drag.

VH-BRF took off from Sydney Harbour once again on 3 October, but this time heading out across the Pacific for Lord Howe Island. It was the start of a long and happy link between *Islander* and her Island.

Islander on the hard standing at Rose Bay after completion of the conversion work.

(Photo via Capt Ron Gillies)

Footnotes Chapter 6.

[1] *To The Ends Of The Air* — G E Livock
[2] *The Defeat Of Distance* — J Gunn (p 332)
[3] *Maritime Is Number Ten* — K C Baff (p 4)
[4] *Bird Of The Islands* — Sir Gordon Taylor (p 13)
[5] *Challenging Horizons* — J Gunn (p 236)
[6] Lord Howe Island Museum — Oral history tapes. Phillip Mathiesen
[7] *Wings To The World* — Sir Hudson Fysh (p 70)
[8] *Aviation Heritage*, Vol 25. No 1. Interview with Stewart Middlemiss by Greg Banfield.

Chapter seven

Islander and the Island

Before looking at *Islander*'s long period of service on the Sydney—Lord Howe Island route, we will take a brief look at the history of the island, and of transport to it. After all, the story of *Islander* is very much a part of the story of 'The Island'.

The history of Lord Howe Island is as old as the history of white settlement in Australia itself. In January 1788, Captain Phillip's First Fleet of ten vessels arrived at Port Jackson (later Sydney) to unload its cargo of convicts and their Royal Marines guards. A few days later one of the vessels, the *Supply*, left for Norfolk Island to found a further colony. En route her commander, Lieutenant Henry Lidgbird Ball, sighted and named Lord Howe Island.

On his return from Norfolk Island, Ball called at Lord Howe Island to collect a quantity of turtles, making the *Supply* the first known vessel to call there. Over the following forty-five years, occasional passing vessels called at the island to replenish food, water, or firewood supplies. Then, in 1833, three families from New Zealand formed the first settlement on the island.

For the remainder of the nineteenth century, the population of the island gradually increased, some families leaving, others moving in. The settlers were largely dependent on barter trade with passing ships for their survival, exchanging island produce for staple commodities they could not grow themselves. Whaling vessels were their most regular callers.

When the Pacific whaling industry started to decline in the 1860s, islanders discovered an indigenous industry to provide for their needs — the sale of palm seeds. There are four unique species of palm found on the island, but one of these, the Howea Palm, became extremely popular in Victorian England, and later Europe. This was due to the species' great tolerance of temperate climates.

Commonly known as the Kentia Palm, it was, by the end of the century, the most common indoor palm throughout the world. From Australia to the Americas and Europe, it became the fashion to give a tropical atmosphere to almost every ballroom and hotel lobby with this spreading palm from Lord Howe Island.

This increase in trade caused Burns Philp, the long established South Pacific trading and shipping company, to commence a regular service to Lord Howe and Norfolk Islands in 1893. This quickly became a good, reliable, and regular service, encouraging early tourists to visit the island.

By the 1930s, this trickle of visitors had increased to such a flow that two island families thought it worthwhile to add rooms to their homes, and establish the first guest houses. These were Ocean View and Pinetrees.

An equable, sub-tropical climate and magnificent scenery were major attractions to visitors. The palm trees and the coral lagoon (the southernmost in the world) reinforced the feeling of being on a tropical island, but the climate was far more pleasant. The family-run guest houses achieved renown for their tables, virtually all produce coming from their own farms, and all baking being homemade. By 1939, there were about 60 beds available for visitors.

The aviation age arrived at Lord Howe Island in 1931 when, completely unannounced, an Englishman arrived one evening in a small single-engined floatplane. The islanders could barely believe it. Their unexpected visitor was Francis Chichester, better known for his post-war exploits as a single-handed yachtsman.[1]

Chichester had flown a Gypsy Moth biplane solo from England to Australia. This gave him a taste for long, trans-oceanic flights, and he started to consider a New Zealand to Australia flight. Like P G Taylor, he recognised that he was lacking in the navigational skills for such a flight, and therefore taught himself astro-navigation and developed his own air sextant.

After converting his Gypsy Moth to a floatplane, he set out from New Zealand in March 1931. The first leg to Norfolk Island went smoothly and without a hitch. The second leg was something of a close call, with both darkness and a nasty storm setting in as he touched down on the lagoon at Lord Howe Island. Islanders rushed to greet him, assisted in mooring the Moth for the night, and he was invited to stay with the Dignam family. Arriving at the lagoon next morning, to set out on the last leg to Australia, Chichester was devastated to find the floatplane lying upside down in the water. The Lord Howe Island lagoon and an overnight storm had taken their first toll of an aircraft.

People quickly gathered and helped to pull the plane from the water. Chichester examined the wreckage ruefully — it was a write-off and the end of his dream. As he considered the possibility of sending the wings to Sydney for rebuilding, an islander said to him "Don't be feeble, man! Why not rebuild the wings yourself?"

Nine weeks of work by Chichester and several islanders saw his plane rebuilt and test flown, and a number of his helpers were taken up for joy rides. Then, with two pigeons aboard for emergencies, Francis Chichester bade a reluctant farewell to the islanders, and successfully completed his flight to Australia.

These pioneering flights, as elsewhere, were inevitably to lead to commercial services. In December 1937 John Burgess, a 29-year-old New Zealander, set out from Southampton in an Imperial Airways Empire class flying boat, the *Centaurus*. His goal was to follow the company's previously surveyed route to Singapore, then survey a flying boat route onward through the Dutch East Indies, Darwin, Sydney, and finally to New Zealand.

John Burgess touched down on the Waitemata Harbour, Auckland, on 27 December. It was his first time home in seven years, and he was greeted with a hero's welcome. Two large flying boats now lay together at their moorings. The previous day Captain Edwin Musick had arrived in Pan American's *Samoan Clipper*, to inaugurate his company's fortnightly San Francisco—Auckland schedule. Sadly, Musick and his crew were killed a week later when his Sikorsky S42 exploded in mid-air near Pago Pago.

Following this survey flight, Imperial Airways' Captain John Burgess said that radio and meteorological services were essential, before scheduled flying between Australia and New Zealand could be contemplated. The service should then be safe and profitable using Shorts S30s, a longer range version of the Empire Boat than his S23 *Centaurus*. [2]

This report led to the expansion of the wireless station which had been set up on Lord Howe Island in 1929 by Amalgamated Wireless of Australia (AWA). Its first operator was Stan Fenton, an islander who had gone off to become a ship's wireless

Lord Howe Island before the flying boats. Pre-war tourists come ashore from the Burns Philp line vessel Morinda. Everything for the island, passengers or freight, had to be brought across the reef by boat.

(Photo Lord Howe Island Museum)

operator, then returned to man the new station. Additionally, the island was designated an emergency landing area for the trans-Tasman flying boats.

With the Second World War came a big expansion in the importance of the wireless station; it became a forward base for the collection of meteorological data, and for monitoring aircraft movements across the Pacific.

The war also brought total disruption to the island's shipping service. Ships were requisitioned for military needs, naval vessels given priority in the dockyards, and often islanders would have to wait months for the next ship to call. To increase this feeling of isolation, the advent of the Pacific War imposed radio silence on shipping movements, and so the island's population had no idea of when to expect their next caller.

This caused a return to the pre-1929 methods of ship watching. Whenever a vessel was expected, a hearty-voiced lookout would be posted on a high point of the island, normally Malabar. On seeing an approaching ship, he would shout "Sail-o". The cry would be picked up and repeated by anyone within earshot, so that it was only a matter of minutes before the entire island was aware of the impending arrival.

Due to the infrequent shipping service, one or two medical emergencies arose in the early period of the war. These were handled by despatching a Catalina flying boat from the RAAF base at Rathmines in New South Wales. Both islanders and the air force saw the advantages of such a means of transport, and so the RAAF began regular flights to the island on a monthly basis.

Any islander needing to get to or from the mainland could travel, emergency needs could be brought over, and the Rathmines Medical Officer and dentist often made the trip to attend anyone in need of their services. The flying boat would normally stay overnight on the lagoon, its crew put up by islanders.

Early flying boat visitor. A Catalina moored in the lagoon at Lord Howe Island.

(Photo Lord Howe Island Museum)

One Catalina captain who made a number of trips to the island at this time was Vic Hodgkinson. He had recently returned to Australia after a tour with Number 10 Squadron on Sunderlands in the UK, and still fondly remembers those flights to Lord Howe Island. "When we arrived all the islanders would come down to the beach — if I recall there were something like a hundred. At that time a large percentage of the men had joined up and were away.

"Once ashore, we were billeted out to the islanders. Of course there were never enough of us to go around. There would normally then be something of a fight for us, with cries of 'No, he is ours, you had someone last time.' That evening there would be a 'Do' in the island hall. Although such events were officially dry, a few bottles would usually appear in the back room."

Vic Hodgkinson was perhaps the first flying boat pilot to experience the hazards of Lord Howe Island weather, in particular the strong downdraughts over the landing area, caused by the proximity of the island's two mountains. Vic describes his experience: "There was a strong westerly blowing as I approached the island, perhaps up to sixty knots. Knowing that there would be very strong downdraughts, I decided to make a low pass to assess the situation."

"I had a few islanders on board, having called at Rose Bay to collect them after coming down from Rathmines. I was probably at two or three hundred feet when the downdraught hit us, and down the old Cat went at a rate of knots. Luckily I was already in the landing configuration, because down she went till she hit the water with a terrific thump. We were down in a stall landing without trying. I got those throttles back and the stick forward pretty quickly, as I certainly didn't want to go up again.

Wartime hospitality. Visiting RAAF aircrew are treated to a picnic by islanders. The vehicle is one of two Thornycroft lorries which were the only motor transport on the island at the time.

(Photo Capt Vic Hodgkinson)

"That night, the gale increased up to ninety or so miles an hour, and she blew ashore. Luckily she fetched up on a sandy beach, and we managed to get her afloat again and fly her back."

Following the war, the air force felt unable to maintain regular flights to the island. Gradually Burns Philp ships came back onto the route, but the company was unable to provide the level of service it had pre-war. The war had taken its toll on the company's fleet, much of which was now very old. Changes in accepted standards of crew accommodation and wages made operations less economic.

As we have seen, by 1947 commercial flying boat services were established, and these now took over the passenger traffic to the island. Not only did the flying boats provide a better transport service for islanders than did the ships, but they allowed a revival and later an expansion of the pre-war tourist trade.

Having now filled in a little on the history of Lord Howe Island, we will return to the period of the nineteen-sixties, and the time when *Islander* was entering operation on the route for Ansett Flying Boat Services.

Perhaps the best way to experience travel to the island at that time is to tap the memories of Frank Chartres, reckoned to be the 'most travelled passenger' on the route. Here he describes a departure from Rose Bay: "Time would come to board, and it was down the ramp and onto the pontoon. The steward would welcome you aboard, having to take your arm as it was quite a step across into the aircraft. So it was all very personal right from the set off. [3]

"Once on board, the seats were so comfortable, and there was so much room. You could stretch your legs out, wander around the flying boat from go to whoa, upstairs

Islander moored on the lagoon. In the background are Mount Lidgbird and Mount Gower.

(Photo Ron Cuskelly)

and downstairs. There were of course meals as appropriate, and they were beautiful meals. Salads, sometimes a hot meal, but mostly salads. Sweets, drinks — with bar service, naturally. Time didn't matter.

"You weren't worried if it was a slow flight, a fast flight, or whatever. Your holiday at Lord Howe Island started and finished at Rose Bay. It was a continuous part of a marvellous experience.

"To some people, one of the problems with the flying boat was that it couldn't get above the weather. In the number of flights I did there were one or two where we had to return to Rose Bay, and one where we had to return to Lord Howe. But as far as I was concerned, the rougher the weather the better I liked it. The biggest disappointment in my life now is that with the present aircraft you fly at 30,000 feet, and miss all the weather.

"In the flying boat it was an experience that was absolutely out of this world. To see storm clouds out in front of the flying boat was the greatest joy of my life. The pilots would try to avoid the biggest of the cumulus clouds, so we would fly through corridors with these clouds towering on each side of us, and as an opening came up we would turn into this or that opening — just like following a laneway between hedges. It was magnificent.

"There came the times when you ran out of any clear weather, and had to go into it. Then it was really rough. But the flying boat was a heavy aircraft, so the roughness was quite tolerable in comparison to being in a small aircraft. It could be likened to the difference between being on a cruiser compared to a dinghy in a rough sea.

"On board the flying boat, you almost felt that you were a member of the crew. Anyone could pop up to the flight deck if they were interested. It made you feel that you were part of an adventure, part of the aircraft. When the meals were over, there was the packing up to be done. I would often slip down and help Lester (Lester Lewis, the longest serving steward on the route) and the hostess with that. Then we would sit and have a drink and a yarn together.

"Flying in, you could usually see the island from 25 or 30 miles out. You would circle the island while the crew took a look at the lagoon. It was like a dream. And then to land on the lagoon: looking out the window to where the bow wave parted, it was like coming up to a sheet of glass. You could see the bottom of the lagoon, the coral — quite magnificent. Then of course you taxied the length of the lagoon, and came to the mooring. The boats would come out to take you ashore from the flying boat.

"Normally we stayed at Leanda Lei, so of course Roy Wilson (the proprietor) was there to meet us. There was no bus, just an old red International table top truck with bench seats on the back. We would be helped up a plank on to the back, the luggage would be piled around us, and off we would go — to be dropped at the various guest houses. This approach, this lack of sophistication, brought a whole new aspect to a holiday. It was this way of life which made Lord Howe Island for me."

So much for a passenger's view of travelling to the island. There seems little doubt that Frank Chartres enjoyed his flights over. For those with a more technical interest in flying, we will take a description of the flight from Captain Lloyd Maundrell, who, incidentally, was to become the most travelled pilot on the route.

"Scheduled flights to Lord Howe Island were dependent upon arriving there about two hours before high tide, and leaving no later than one hour after the tide. The reason being that at low tide, only two to three feet of water covered the island lagoon. As the draught of the flying boat was four feet nine to five feet, insufficient water was available for our operation.

"In order to meet these requirements, night flying was essential, although this was not permitted at the Lord Howe end because of the mountainous terrain. The trip was approximately three hours, outgoing at seven to eight thousand feet, and inbound at two to four thousand feet — in order to utilise the prevailing winds. These normally tended to be westerly.

"Operation of the flying boats at the island was most demanding. Heavy seas sometimes crossed the reef, causing a nasty three to five foot swell in the lagoon. Take-offs and landings could become rather hazardous as the mountainous terrain, gusty high winds, and severe turbulence could be quite frightening at times. Because of these conditions, the lowest safe altitude for a radio let-down was four thousand feet.

"Navigational aids originally were restricted to HF for communications, and an ADF beacon. We used astro-navigation early on, but eventually we got DME (distance measuring equipment) and VHF for communications — so we then scrapped the astro altogether."

Islander washed ashore at Windy Point, May 1965.

(Photo Lord Howe Island Museum)

It was, in fact, the opinion of most pilots flying the route that the operation at the Lord Howe end was so marginal, that in normal circumstances the Department of Civil Aviation would never have allowed it. They tolerated it only because there was no other practical form of transport available to the island at that time.

It is also interesting to note that, as the years passed, and many pilots came and went as first officers, no new captains were ever introduced on the route. Right to the end of the service it was the same small group of 'Old Stagers' who were the captains. All had either been wartime flying boat pilots, or had come to flying boats within a few years of the end of the war.

Despite the demanding conditions, the safety record of the service was exemplary. Not a single passenger was ever injured during 27 years of the flying boat service. The two cabin crew were once slightly injured in *Islander* during severe turbulence, because they, unlike the passengers at the time, were not strapped in.

Islander had been operating on the Lord Howe Island route for barely eight months when near disaster struck, and she almost suffered the same fate as had *Pacific Chieftain*. She had flown out to the island on 31 May 1965, on a scheduled service. Having arrived at the island, a forecast of strong winds caused a decision to overnight with the boat — not too unusual, especially at this time of year.

It was during winter, particularly the late winter months, when the depressions of the lower latitudes move further north, that the island was often hit by powerful westerlies. This particular night was no exception, with the wind getting up to cyclone strength.

Mick Nichols was heading up to start duty at the Met Station, around midnight, when he sighted *Islander* up on the beach at Windy Point — exactly where *Pacific Chieftain* had come to grief, two years earlier. To raise the alarm he made his way to the Bowling Club — the social centre of the island — where Jim Dorman was running a dance. Jim recalls the night clearly: [4]

Salvage work begins. Fuel has been emptied into drums and volunteers sit on the port wing as others attempt to raise the starboard wing and support it with sandbags.

(Photo Lord Howe Island Museum)

"I stopped the dance, obviously, and called for volunteers. Quite a number, both locals and tourists, volunteered. First we had to get vehicles and torches to illuminate the beach, as it was a pitch-black night. When we got there, and saw the huge sea running, there were not so many volunteers who would actually wade into the water and do the real work.

"The pilots had been woken, and had come down from the Blue Lagoon. One of the floats had broken off and she was laying over on one wing. The pilots' main concern was to try to raise that wing by sandbagging, to get the strain off it. As she was sitting in soft sand, they did not consider that there would be any real damage to the hull."

Having learnt from the experience of *Pacific Chieftain*, no attempt was made to refloat *Islander*. Instead, work concentrated on trying to lighten her, by unloading her fuel into 44 gallon drums, and to get her further back up the beach, clear of damage by the sea.

The island's road grader was brought down, and helped haul *Islander* up the beach, as the sea pounded her back. When she was as far back as possible, a boat was used to lay anchors out ahead of her, to try and hold her head into sea.

The only good news about these beachings at Lord Howe Island was that the dangerous winds were the westerlies, and at least these always fetched the flying boats up on the sandy beach, which gave them a chance of survival. Had they been carried onto the reef, there would have been no hope of salvage.

Jim Dorman continues: "We got a lot of people sitting on the opposite wing to try to get the thing counterbalanced, but all this was happening at night in very, very strong winds and a heavy sea. As you can imagine, it wasn't altogether what you would call a pleasant job. Those of us below the waterline had the odd big wave breaking over the top of us."

108 · The Last FLYING BOAT

A replacement float has been shipped out from Sydney and fitted to the starboard wing.
(Photo Lord Howe Island Museum)

A channel is dug around the grounded flying boat and sandbagged to prevent collapse.
(Photo Lord Howe Island Museum)

Two launches are attached to the bow, ready for the attempted refloating at high tide.
(Photo Lord Howe Island Museum)

In fact Jim lost his glasses in one such wave, and luckily a sharp-eyed tourist with a torch managed to locate them immediately. As the night wore on, the number of volunteers on the job dwindled, but by daybreak the small team remaining had *Islander* temporarily secure, high and dry up the beach.

Islanders, knowing that they were totally dependent on the flying boats, were always willing helpers in emergencies such as this. Despite this, Ansett announced the following morning that they would pay wages to those employed full-time on the salvage work.

Now came the problem of how to carry out temporary repairs. In a more usual location, this would not have been a great problem, but the island was no normal location. Men and materials had to be got to Lord Howe, but by what transport? One of the flying boats was laid up each year during the winter 'off' season, for its major two-yearly overhaul and inspection. Luck had it that *Beachcomber* was out of commission now.

At least the BP line vessel *Tulagi* was due out of Sydney soon on its regular run to Norfolk Island, so four engineers from Rose Bay and the necessary tools and materials for temporary repairs sailed with her. This included a replacement float.

They arrived at Lord Howe Island on Sunday 6 June, where the westerlies were still too strong to allow them to land over the reef. Instead, the men and their gear had to be landed on the opposite side of the island, at Ned's Beach. With the wind blowing at over 40 knots, they were also prevented from making an immediate start on their work. *Islander* had by now been on the beach a week.

Repairs weren't the only problem facing Airlines of NSW. There was also the problem of stranded passengers. The forty or so who had been due to fly back with *Islander* were now a week overdue. Added to that, the tourists who had flown out on her had now reached the end of their stay, so the airline was meeting the accommodation bill for eighty stranded passengers. For the lucky few with no pressing commitments in Australia, it was quite a bonus.

In an attempt to get his passengers back, Stewart Middlemiss had diverted the P & O-Orient liner *Orsova* to the island. She was on passage from Sydney to Auckland at the time, and could have taken the stranded passengers on to New Zealand, from where they would have been flown back to Sydney. However, when she arrived off Lord Howe Island on the Saturday, conditions were so bad that it was too dangerous to attempt getting the tourists to her by boat. More expense for Ansetts, and to no avail.

The repair team, assisted by a number of islanders, was now making progress. The starboard float had been refitted and rigged, and a broken section of leading edge fitted back onto the wing. The wing tip still looked horribly crumpled, but was considered quite safe for a flight back to Sydney. Now came the problem of refloating her.

A channel was dug around her, sandbagged to prevent it from collapsing, and linked to the sea. This now gave two to three feet of water around the hull (she draws nearly five feet). Two sturdy island boats, the *Alena* and the *Chieftain*, were put on long tow lines to *Islander's* bow. Groups of men were organised under the wingtips, holding poles padded at the top. These were to hold the aircraft level, so neither float would touch the sand and create drag.

The sand flies and trees bend as Islander moves back into the lagoon. Locals and tourists support the wings, two island boats haul on the bow and Captain Bill Wilcher applies full power on all four engines.

(Photo Lord Howe Island Museum)

At high tide all was ready for the attempt. The two boats took up the tow, and Capt Bill Wilcher opened *Islander's* four engines up to full power. The five thousand horsepower threw up a vicious spray of sand and water behind the flying boat. They rocked her a little to break the suction, and *Islander* moved forward. She was back in her element, the sea. After a short test flight she returned to Sydney on the morning of Thursday 10th, crew only on board. [5]

At this stage, consideration was being given to the chartering of an RNZAF Sunderland, as being the only feasible method of getting the stranded passengers off the island. But the maintenance staff worked feverishly at Rose Bay, mainly to fabricate new brackets for attaching the wing leading edge, as well as a thorough inspection of possibly stressed components. By Friday 11th *Islander* was airworthy again, and returned to the island to bring back the first load of stranded passengers.

The following day, she flew out again to retrieve the last of the 'castaways', and the crisis was over. Plans to charter an RNZAF Sunderland did not have to be put into effect. But the total cost of the incident to Airlines of NSW was considerable, and it certainly illustrated the type of problems they could be faced with, through having to maintain the flying boat service.

Once again, therefore, the perennial issue of 'replacing the flying boats' came to the fore. Back in mid-1962, long before *Islander* had arrived in Australia, the question of building an airstrip on Lord Howe Island first received serious attention. Airlines of NSW had stated publicly that the flying boats were near the end of their economic life, and that there were no modern aircraft available as replacements. Even if a suitable replacement could be found, the capital expenditure could not be justified if it were only to fly the one route, with a frequency of one flight per week in winter and a maximum of five per week in the season.

David Murray, Ansett's station manager at Lord Howe Island, stands in the door of the island's terminal building, by the lagoon.

(Photo Lord Howe Island Museum)

In addition to this view from Ansetts, there was a strong local lobby developing in the Rose Bay area, to push for the closing of the flying boat base. It was hard to see how residents could complain of excessive disturbance from an average of two to three flights per week, especially as the noise of a Sandringham passing directly overhead was barely audible in comparison to a jet. However, it must be admitted that the occasional flight was at a most un-neighbourly hour. Rose Bay departures could be as early as 2.30 am, to allow arrival at first light on the lagoon and fit in with the state of tide there.

The upshot of all these discussions was that a Department of Civil Aviation team, led by their senior airport engineer, was despatched to the island. Ansett had stated that a service would not be viable unless an aircraft the size of a Fokker Friendship could use any proposed airstrip. The team returned from Lord Howe to report that a strip could certainly be built, but because of the terrain of the island, any strip would have to be partly built on land reclaimed from the lagoon.

Ansett Transport Industries offered to pay a substantial part of the cost, but arguments developed between Commonwealth and NSW State governments over responsibility for the remainder. Gradually the discussions died away and nothing was done. Despite *Islander's* 1965 mishap, yet again it wasn't long before the matter was forgotten.

With *Islander* back on the route, life was soon back to its normal quiet pace for the 250 islanders and their steady flow of tourists. About half of the island's work-force was dependent, directly or indirectly, on the tourist industry. Several island residents, however, were even more directly involved with the flying boats for their livelihood...

Longest established of these was Harry Woolnough. Harry had started life as a marine engineer, but was put out of work by the Great Depression. By chance, he was offered a position as crew on a yacht and readily accepted it. Thus, in 1931, he found himself visiting Lord Howe Island with the yacht *Spumedrift*. He thought the island would be a most attractive place to spend some time, and was highly delighted to find that his skills were in demand there.

From carrying out work for islanders, he went on be appointed by the Department of Civil Aviation to maintain their flying boat moorings and other facilities on the island. There were of course no civil flying boat services to the island at that time (1942) — purely an emergency landing area for the trans-Tasman route. This had come as a result of the report made by John Burgess, following his route-proving flight in the Empire Boat *Centaurus*.

Once the post-war flying boat services commenced, Harry Woolnough's job assumed a greater importance. Not only were his moorings in regular use, but he had to attend the arrival and departure of each flight. A DCA launch was provided at the island for this duty.

Harry's first task in preparation for an arriving flying boat was to check that its mooring was prepared. The mooring pendant would be attached to the buoy, and left ready for pickup by the steward with his boat hook. Next, a number of buoys with red flags attached had to be positioned over several shoals — these could rip the bottom out of a Sandringham if hit on a landing run.

With the danger markers safely in place, a sweep then had to be made to check for any floating debris on the lagoon. An unnoticed log or heavy baulk of timber was likewise a major hazard. Once satisfied that everything was prepared, Harry's next responsibility was to radio "Landing Instructions" to the approaching aircraft. Its captain needed to know the direction and strength of the wind, sea state, and which mooring to use. The DCA launch was, after all, the equivalent of the control tower at a land airport.

David Murray was another person whose job on the island was directly dependent on the flying boats. He had been appointed traffic manager on Lord Howe Island by Airlines of NSW, back in 1957. Looking after the passengers' needs at this destination would not seem a very demanding job. After all, the absolute maximum number of flights could be two a day with forty or so passengers on each, and to be this busy was rare. [6]

But the traffic manager on Lord Howe Island had some unusual tasks. The island's communication system, like its transport system, was from an earlier age. Telephones were not installed around the island until 1982, long after the departure of the flying boats. Thus, whenever there was a delayed flight, unscheduled arrival, and so on, David Murray had to jump on his bicycle (in later years a motor scooter) and do the rounds of the island guest houses, to get the news through. This could be at any hour of the day or night.

On the subject of communications, the Morse system which was first operated by Stan Fenton back in 1929 did not go out of use until 1975, by which time it was the last Morse system still used in Australia for public correspondence. At times, such as a delayed return flight, the harassed operator was kept busy tapping out hundreds of messages. Stan Fenton himself did not retire as officer in charge of the communications station until 1968.

As is by now fairly apparent, life on Lord Howe Island in the flying boat years did not proceed at a breakneck pace, nor did islanders show great enthusiasm for embracing the latest fads and fashions. This, of course, is what gave the island its charm and attraction to visitors.

The accommodation was generally full board at the few guest houses, and they tended to take full responsibility for their guests' well-being and entertainment. They would take turns at borrowing the Lord Howe Island Board's truck, and give their guests a conducted tour of the island. Another day might be spent on a barbecue picnic, another on a fishing expedition (most guest houses operated their own launch). Evenings were usually spent with self-made entertainment such as singsongs.

Arrival and departure of the two flying boats, although commonplace, was always looked upon as being something of an event, and would attract a good crowd of spectators. We will turn again to Frank Chartres for a description. Frank, incidentally, made a total of 36 trips to the island by flying boat.

"The mere arrival or departure of a flying boat was an island event. You had forty-odd people all leaving at the same time. You had all made friends on the island, so of course all the tourists and most of the locals would be down at the jetty to see them off — and to see what sort of talent was on the incoming flight. Those leaving would be presented with leis made out of hibiscus. It was tradition that these be thrown into the water as you went out to the plane. If the lei was washed onto the beach it was an omen that you would return to the island. Naturally they always did, there wasn't anywhere much else that they could wash to."

Although the 1962 discussions on the building of an airstrip had gradually died away, the topic was making the news again in the early 70s. In 1966, Sir Reginald Ansett had notified the Commonwealth and NSW governments that the flying boats could not remain in service beyond 1970. Both took little or no notice of his pronouncements, and he extended his deadline, first to November 1973, and then to May 1974. Finally, when at the last moment the two governments realised that they had a responsibility to provide the islanders with some form of transport, the Australian Army was brought in to build a small airstrip. It was a truly last minute solution.

All of this was announced in a Department of Civil Aviation (by now known as the Department of Transport) report of February 1974. Four reasons were given for ending the flying boat services. They were: (a) Age and maintenance costs of the Sandringhams. (b) Intending retirement of the experienced captains. (c) Increased hazards of shipping and high-rise developments on and around Sydney Harbour.
(d) The high cost of subsidising the service and the operating bases at Rose Bay, Lord Howe, and Redland Bay, Brisbane (the bad weather alternative to Rose Bay).

Only when the flying boats were on the point of disappearing forever from the lagoon at Lord Howe Island, was their true value to the community fully realised. It had always been obvious that they provided an essential transport service. But now that the alternatives had to be seriously assessed, it became apparent that nothing else really suited the island's needs.

Being only seven miles long by one mile in width, and mainly mountainous, the island had precious little suitable land to lose to an airstrip. The original suggestion had been for a strip along the edge of the lagoon, but a geological survey proved this impracticable. The only alternative seemed to be a strip slashing across the narrow, flat, centre section of the island, effectively cutting it in two. Islanders who had seen airport construction create a great scar across other islands did not want this.

Captain Bill Wilcher at the controls of Islander.

(Photo Harry Williams)

Then there was the subject of size. The guest house proprietors thought it necessary to be able to land planes of similar capacity to the flying boats, say Fokker Friendships. But if you could get in one Fokker a day, you could get in twenty. No one wanted the island ruined by being turned into another Queensland Gold Coast. Also, to take a Fokker, the strip would have to be extended out into the lagoon — environmentally unacceptable to most.

So the debate became quite intense, both on the island and on the mainland. Islanders were divided in their opinions. Roy Wilson, proprietor of Frank Chartres' favourite guest house, thought that a strip cutting across the centre of the island would cut its heart out. "If that is our only choice, I'd rather have no air service, even if it means the end of my business. It amounts to this: We could have an island with no airstrip, or an airstrip with no real island. I want an island."

Phil Dignam, the islanders' elected representative on the Lord Howe Island Board, was of similar opinion. "Islanders are prepared to accept an airstrip on the original site, but I doubt very much whether the majority would accept one running right across the middle of the island."

The opposite point of view was put by tour boat operator Clive Wilson. "If we don't have an air service, the tourists can't come, and all our investments will be valueless."

All that the flying boats had ever required from the island was use of the lagoon. They had met the islanders' own transport needs, as well as bringing in up to 5,000 visitors a year. Plenty to keep the six guest lodges going, but they couldn't have brought more tourists at peak season had it been asked. So they were adequate, posed no threat of overdevelopment, and left the island's natural beauty totally unspoilt. But this perfect solution was soon to be a part of history.

Back in Sydney, Bryan Monkton raised an almost lone voice in suggesting a different solution. Certainly, he agreed, it was not possible to buy a new Sandringham, but there were still other marine aircraft available. He suggested the use of one of the American-built Grumman amphibians for the route. This could still use the lagoon at Lord Howe, but operate from Sydney's land airport at the other end. At least the Rose Bay residents would be happy!

However, the die was already cast. Such a service would still require a subsidy, and the government was now determined to build an airstrip and scrap the subsidy. In March 1974, landing craft arrived at the island, and off rolled the first of the khaki-painted earth movers; the army had invaded! Quickly they set to work on the construction of an airstrip, which did cut right across the heart of the island.

Now, it seemed, no one was happy with the solution. Environmentalists, particularly those from the mainland, were up in arms. Islanders were dissatisfied, mainly because the strip would only be 3,000 feet long. This would only allow less commercially-viable, light twin-engined aircraft to use it. The Fokker Friendships required 5,000 feet. There was even talk of the likelihood of the tourist facilities becoming worthless, and most of the population having to abandon their properties and leave for the mainland.

The flying boat captains joined others in their dire predictions for the future. They could see that winds gusting around the island's two mountains, almost on the edge of the airstrip, would be a real hazard to light aircraft. Lloyd Maundrell commented: "Unless they have the right type of aeroplanes and, incidentally, the right type of pilots who can handle those conditions, I don't think they will have a really good service."

Only the long complaining Rose Bay residents had cause for celebration. One good lady summed up her thoughts, "You know, I will really be very glad when these noisy things go. After all, you pay an awful lot of money for a house in Rose Bay, and you might as well live in Mascot." (Sydney's land airport). No doubt in later years, the same lady would lament to her guests "How terribly sad that those romantic flying boats have departed. It was so wonderful to watch them skimming gracefully across the harbour."

Although work on the airstrip was progressing, time was running out too quickly. Airlines of NSW had scheduled the last flying boat service to the island for 31 May, and in fact both boats had already been sold to an American airline, which was anxious to put them to use in the Caribbean. With no chance of the airstrip being ready until September, the Commonwealth Government obtained yet another extension of the flying boat service by Ansett. This was to be a "communication link" for essential traffic only, a large proportion of which was connected with the building of the new airstrip.

Islander, however, would not be involved in this extension of the service. On 30 May she made what was scheduled to be her last flight to the island. On return to Rose Bay it was up the slipway for her, as there was work to be carried out in preparation for her new owners. It was now down to *Beachcomber* to maintain the final few months of the old service, and to be present at the farewell ceremonies when the long-threatened end finally came.

Unpredictable to the end, the old Sandringhams were soon to disrupt the best laid plans of governments and airlines. The following weekend *Beachcomber* was over at the island, on one of her official charters. The weather blew up on the Sunday afternoon, so she was forced to overnight. And, of course, the inevitable happened. Monday morning saw her on the beach.

AIRLINES OF N.S.W. Timetable

7 LORD HOWE ISLAND

	Dep. Sydney	Arr. L.H. Is.	Dep. L.H. Is.	Arr. Sydney		Dep. Sydney	Arr. L.H. Is.	Dep. L.H. Is.	Arr. Sydney
	Flight No. 860		Flight No. 861			Flight No. 860		Flight No. 861	
OCTOBER, 1973					Wed. 19th	12 noon	3.50 p.m.	4.30 p.m.	7.20 p.m.
Thu. 18th	7.00 a.m.	10.50 a.m.	11.30 a.m.	2.20 p.m.	Thu. 20th	1.00 p.m.	4.50 p.m.	5.30 p.m.	8.20 p.m.
Mon. 22nd	12 noon	3.50 p.m.	4.30 p.m.	7.20 p.m.	Fri. 21st	2.00 p.m.	5.50 p.m.	6.30 p.m.	9.20 p.m.
Thu. 25th	1.30 a.m.	5.20 a.m.	6.00 a.m.	8.50 a.m.	Mon. 24th	3.00 a.m.	6.50 a.m.	7.30 a.m.	10.20 a.m.
Tue. 30th	6.00 a.m.	9.50 a.m.	10.30 a.m.	1.20 p.m.	Thu. 27th	5.00 a.m.	8.50 a.m.	9.30 a.m.	12.20 p.m.
NOVEMBER, 1973					Fri. 28th	5.30 a.m.	9.20 a.m.	10.00 a.m.	12.50 p.m.
Fri. 2nd	8.00 a.m.	11.50 a.m.	12.30 p.m.	3.20 p.m.	Mon. 31st	7.30 a.m.	11.20 a.m.	12 noon	2.50 p.m.
Mon. 5th	11.00 a.m.	2.50 p.m.	3.30 p.m.	6.20 p.m.	**JANUARY, 1974**				
Tue. 6th	12.00 p.m.	3.50 p.m.	4.30 p.m.	7.20 p.m.	Wed. 2nd	9.30 a.m.	1.20 p.m.	2.00 p.m.	4.50 p.m.
Fri. 9th	2.00 p.m.	5.50 p.m.	6.30 p.m.	9.20 p.m.	Thu. 3rd	10.30 a.m.	2.20 p.m.	3.00 p.m.	5.50 p.m.
Tue. 13th	4.30 a.m.	8.20 a.m.	9.00 a.m.	11.50 a.m.	Fri. 4th	12.00 noon	3.50 p.m.	4.50 p.m.	7.20 p.m.
Thu. 15th	6.30 a.m.	10.20 a.m.	11.00 a.m.	1.50 p.m.	Mon. 7th	2.00 a.m.	5.50 a.m.	6.30 a.m.	9.20 a.m.
Fri. 16th	8.00 a.m.	11.50 a.m.	12.30 p.m.	3.20 p.m.	Tue. 8th	2.30 a.m.	6.20 a.m.	7.00 a.m.	9.50 a.m.
Mon. 19th	11.30 a.m.	3.20 p.m.	4.00 p.m.	6.50 p.m.	Wed. 9th	3.30 a.m.	7.20 a.m.	8.00 a.m.	10.50 a.m.
Fri. 23rd	2.30 a.m.	6.20 a.m.	7.00 a.m.	9.50 a.m.	Thu. 10th	4.00 a.m.	7.50 a.m.	8.30 a.m.	11.20 a.m.
Tue. 27th	4.30 a.m.	8.20 a.m.	9.00 a.m.	11.50 a.m.	Fri. 11th	5.00 a.m.	8.50 a.m.	9.30 a.m.	12.20 p.m.
Thu. 29th	6.00 a.m.	9.50 a.m.	10.30 a.m.	1.20 p.m.	Mon. 14th	8.00 a.m.	11.50 a.m.	12.30 p.m.	3.20 p.m.
Fri. 30th	7.00 a.m.	10.50 a.m.	11.30 a.m.	2.20 p.m.	Tue. 15th	9.30 a.m.	1.20 p.m.	2.00 p.m.	4.50 p.m.
DECEMBER, 1973					Wed. 16th	10.30 a.m.	2.20 p.m.	3.00 p.m.	5.50 p.m.
Tue. 4th	10.00 a.m.	1.50 p.m.	2.30 p.m.	5.20 p.m.	Thu. 17th	12 noon	3.50 p.m.	4.30 p.m.	7.20 p.m.
Thu. 6th	12.30 p.m.	4.20 p.m.	5.00 p.m.	7.50 p.m.	Fri. 18th	1.00 p.m.	4.50 p.m.	5.30 p.m.	8.20 p.m.
Mon. 10th	3.00 a.m.	6.50 a.m.	7.30 a.m.	10.20 a.m.	Mon. 21st	2.30 a.m.	6.20 a.m.	7.00 a.m.	9.50 a.m.
Tue. 11th	3.30 a.m.	7.20 a.m.	8.00 a.m.	10.50 a.m.	Tue. 22nd	3.00 a.m.	6.50 a.m.	7.30 a.m.	10.20 a.m.
Wed. 12th	4.30 a.m.	8.20 a.m.	9.00 a.m.	11.50 a.m.	Wed. 23rd	3.30 a.m.	7.20 a.m.	8.00 a.m.	10.50 a.m.
Fri. 14th	6.30 a.m.	10.20 a.m.	11.00 a.m.	1.50 p.m.	Thu. 24th	4.00 a.m.	7.50 a.m.	8.30 a.m.	11.20 a.m.
Mon. 17th	10.00 a.m.	1.50 p.m.	2.30 p.m.	5.20 p.m.	Fri. 25th	4.30 a.m.	8.20 a.m.	9.00 a.m.	11.50 a.m.
Tue. 18th	11.00 a.m.	2.50 p.m.	3.30 p.m.	6.20 p.m.	Tue. 29th	7.00 a.m.	10.50 a.m.	11.30 a.m.	2.20 p.m.
					Wed. 30th	8.00 a.m.	11.50 a.m.	12.30 p.m.	3.20 p.m.

GENERAL INFORMATION

AIRPORT DEPARTURE REQUIREMENTS

Passengers travelling privately to the airport must arrive AT LEAST 15 MINUTES BEFORE SCHEDULED DEPARTURE TIME.

YOUR TICKET: You may buy your Flight Ticket at all recognised Travel Agencies or direct from any Airlines of N.S.W., Ansett Airlines of Australia, or Airlines of South Australia offices. It is essential that you collect your ticket as soon as possible after your booking has been made. Naturally, in the interests of other intending passengers we cannot hold unticketed seats indefinitely.

GROUND TRANSPORT: Transport is available at a nominal rate between Company offices and airports.

CANCELLATION FEES: The full rate will be refunded if more than 24 hours notice is given; if not less than 12 hours or more than 24 hours notice, less 10 per cent; if not less than 6 hours or more than 12 hours notice less 50 per cent; under 6 hours, no refund unless authorised by Head Office.

CHILDREN: Children under 3 years each accompanied by an adult travel FREE. Children 3 years of age and under 15, or children under 3, occupying seats — HALF FARE. Children under 5 years must be accompanied by an adult but Airlines of N.S.W. will be pleased to provide an Air Hostess to act as a companion if so desired. A single fare only is charged for the companion and half fare for the child.

STUDENTS' CONCESSIONS: Students enrolled for a FULL-TIME DAY COURSE during THE CURRENT SCHOLASTIC YEAR and NOT in receipt of any REMUNERATION will be granted a concession as follows:—

GROUP "A" — Students 15 years of age and under 19 years enrolled at an Educational Establishment, but excluding a University or College of Advanced Education — 50 per cent concession (half adult fare). Students in Group "A" may obtain a concession on production of a certificate from their Principal or Headmaster.

GROUP "B" — Students who are under 26 years attending a University, College of Advanced Education (as prescribed in the States Grants (Advanced Education) Act 1969), Theological College, or Teacher College Students — 25 per cent concession (three-quarter adult fare).

University students will, on production of their certified identification card, be granted their concession. No claims for a retrospective adjustment will be entertained.

Students' concessions are not available on Lord Howe Island Services.

BAGGAGE: Each ticket holder travelling solely within Australia is allowed free carriage of one piece of baggage with maximum linear dimensions (i.e. length + width + depth) up to 140 cm.
Subject to space availability, additional pieces will be carried and excess baggage charges will be payable as follows:—
(a) On the first two pieces of excess baggages as per rate hereunder:
(b) On third and additional pieces of excess baggage at three times the basic rates hereunder:
$1 for a journey where the first class fare is no more than $35; $2 between $35 and $60; $3 between $60 and $90; $4 between $90 and $120; $5 between $120 and $200; $6 over $200.
International passengers using ANSETT Airlines of N.S.W. connecting services are entitled to a similar free baggage allowance as permitted by the Overseas Airline, i.e. 30 Kilos First Class; 20 Kilos Economy Class.

FREIGHT: Airlines of N.S.W. carries air cargo on all services. Air cargo rates are extremely low, and compare more than favourably with surface transport. For further information contact your nearest Airlines of N.S.W. Office.

GROUP TRAVEL: Special concession for group travel — 10 per cent, 15 or more passengers.

ONWARD TRAVEL: No matter where your final destination may be — within Australia or Overseas — Airlines of N.S.W. will gladly arrange your onward itinerary by land, sea or air.

AVIS CARS: Are available at principal townships. Enquire at your travel agent for details.

The Lord Howe Island timetable, October 1973 — January 1974

Passengers and crew coming ashore from Islander at Lord Howe Island. The steward is still in the bow tending his mooring gear.

(Photo via Bruce Robertson)

Ironically, the passengers she had been due to take back to the mainland were a group of conservationists, who had come over to study the effects that the airstrip construction was having on the island.

Not only was *Islander* up the slipway for maintenance, on Monday morning an American crew had arrived in Sydney to prepare for her flight to the Caribbean. Inspection of *Beachcomber* out on the island showed her to be quite badly damaged, more so than had been *Islander* in 1965, but she was probably still salvageable. Repairs were expected to take several months.

Passengers marooned on the island again, no island service for mail and emergencies — it was back to old days. There was nothing for it but to put off the disappointed Americans, and prepare *Islander* to go over to the island once again. Thursday 13 June saw her out over the Pacific on the much-travelled three hour haul to Lord Howe, then touching down on the lagoon under the gaze of a very relieved party of conservationists.

Islander was to make nine more round trips to the island on the extended service, before making her real 'final flight' on 15 August 1974. By now, the airstrip was close to completion and *Beachcomber* was back in service, so she could be left to handle the remaining flights. The airstrip had already seen its first landing when, on 5 August, an RAAF Caribou transport plane had touched down in strong winds on the partly-completed strip. This emergency mission was to take out a severely injured islander, and the air force pilot later described the flight and landing as "harrowing".

Three weeks after *Islander's* final trip, a little group gathered at the Rose Bay terminal to see *Beachcomber* off on the last official flight to the island. She left at 8.30 am on Tuesday 10 September, fully booked, although Frank Chartres had ensured that he was amongst the passengers. Fittingly, Lloyd Maundrell was at the controls, on his 994th flight to the island. "I would have liked to make the full one thousand", he said.

Later he was philosophical about the ending of the service. "It is very upsetting, but all good things must come to an end. I've brought the Lord Howe Islanders' kids back and forth for 22 years. I've seen them grow up. The island has been so dependent on the service, its been a lifeline for them. We've carried cattle, sheep, chooks, dogs. We've never harmed a passenger and we've never lost a mail bag."

Beachcomber returned to the island the following day to lift out assorted company equipment. And so, after 27 years, the chapter of flying boat services to Lord Howe Island was finally closed. As we have dealt so closely with island life over those years, it is perhaps worth adding a short postscript here, on how the island has fared since the service ended.

The most obvious change, of course, is the great ugly scar across the island. But visitor numbers have remained high enough to satisfy the guest houses, and overdevelopment has not occurred.

When, in December 1982, Lord Howe Island was granted World Heritage Listing, any fears of despoilment through overdevelopment were permanently ruled out.

The new air service, although it has been run by various operators through the intervening years, has proved more satisfactory than many had predicted. The first change was an immediate fare rise from $108 on the flying boats to $130. It now took four flights to bring in as many passengers as could one flying boat. But the small aircraft handled the island's weather conditions better than expected, and have suffered less from weather delays than did the boats.

We will give Frank Chartres the last word on travel to Lord Howe Island. "Nowadays you put up with the air trip because you like the island. The planes are terribly cramped, they are simply a means of getting from A to B, with the minimum of comfort. In the days of the flying boats you went as much for the flight over as for the island. It was the most perfect marriage of location and a means of getting to it."

Footnotes Chapter 7.

[1] *The Lonely Sea And The Sky* — Francis Chichester (p 142)
[2] *Flight Path South Pacific* — Ian Driscoll (p 70)
[3] Lord Howe Island Museum — Oral History Tapes, Frank Chartres.
[4] Lord Howe Island Museum — Oral History Tapes, Jim Dorman.
[5] Lord Howe Island Museum — Oral History Tapes, Norman Simpson.
[6] Lord Howe Island Museum — Oral History Tapes, David Murray.

Chapter eight

The Caribbean

Whilst from the mid-1950s onward, the primary reason for the existence of Ansett Flying Boat Services was to maintain the Lord Howe Island services, these were certainly not the only flights they operated. Over the winter months there were regular South Pacific cruises although by 1964, when *Islander* was brought into service, these were almost a thing of the past.

In its heyday, Rose Bay (or Sydney Water Airport) had been a hive of activity; in fact it was Australia's premier overseas air terminal. In the early post-war years, there would be BOAC flying boats coming in from England, TEAL boats which operated across the Tasman, not to mention the numerous Qantas and Trans-Oceanic flights to other Australian cities, and destinations throughout the Pacific.

All of this necessitated quite an infrastructure. Apart from the terminal building there were two large hangars, the slipway, a control tower, Braby pontoons for loading passengers and freight, and a small fleet of launches. Southampton was the only other port on the England—Australia route with such extensive facilities.

By the time Ansett moved their flying boat operations to Rose Bay in 1953, the importance of the base was rapidly dwindling. One of the hangars had been moved to Mascot, and other facilities had been scaled down. They were, however, left with a very well equipped base for their relatively small operation.

In an attempt to find some additional employment for their two flying boats in the sixties and early seventies, Ansett undertook occasional charter work. One regular destination for such flights was Lake Eucumbene, in the Snowy Mountains of south eastern New South Wales.

Lake Eucumbene is a vast man-made lake, formed during the construction of the Snowy Mountains Hydro Electric Scheme. A flight there involved passing over scenery similar to that previously seen by *Islander* in her Norwegian days — rugged, snow-covered peaks. A charter flight to Lake Eucumbene was usually a weekend affair, with the boat remaining on the lake, and then returning to Sydney on the Sunday. Regular customers were fishing parties and aviation enthusiast groups.

Beachcomber and *Islander* were also chartered from time to time by land development agents. They would fly prospective customers to development sites at Smith Lake and Wallis Lake on the coast, about one hour's flying time north of Sydney. The return trip was always a source of amusement to the steward and hostess. Passengers who had signed a deal with the agents were bought drinks and generally fussed over, while the non-purchasers were totally ignored.

There were still the occasional Pacific island tours during *Islander's* time, but these were all charters. Once a year, Sir Reginald Ansett would use the boat to take a group of corporate guests away for a few days, sometimes to the Barrier Reef, sometimes the Pacific.

One such trip was a big-game fishing expedition to New Zealand's beautiful Bay of Islands. For these business charters, the usual upper cabin seating was removed to create a large bar-lounge. Sometimes even a poker machine (an Australian

Islander passing over the Snowy Mountains en route to Lake Eucumbene.

(Photo Harry Williams)

version of the one-armed bandit) was fitted, in keeping with the prime attraction of Sydney club life of those days.

Sir Reginald Myles Ansett (RM or The Boss) would normally take his family on holiday twice a year in one of the flying boats — although he always travelled on a scheduled flight. His wife and children would normally be given Cabin A (the smaller forward cabin), where no doubt the cabin staff gave them excellent attention. Meanwhile RM would settle down into the jump seat on the flight deck, invariably accompanied by a couple of paperback novels, a bottle of the best Scotch, and one glass.

During the latter years of the flying boat operation, crew members of international airliners calling at Sydney became regular visitors to Rose Bay. Because of its uniqueness, the operation was of interest to them, particularly those who had "served their time" on flying boats elsewhere in the world. On one occasion a charter flight was even put on for visiting BOAC aircrew.

One such visitor to show even more interest than most was a Pan Am captain by the name of Charles Blair. Captain Blair had flown into Sydney in 1968, and Ron Gillies took him and his wife for a trip around the harbour while carrying out a test flight in one of the Sandringhams. If ever a man had flying boats in his blood, it was Charles Blair.

Back in 1938, a new airline — American Export Airlines — was being set up to challenge Pan American on the Atlantic route. Charles Blair was appointed their senior pilot, and was responsible for bringing into service a new type of flying boat, the Sikorsky VS44, right through from the test flying of the new boat. Then he had to select aircrews, all landplane pilots, and train them to a stage of competence to make American Export a viable competitor to Pan American. This done, he flew the Atlantic route throughout the war. [1]

Arriving at Waitangi in the Bay of Islands, New Zealand, on a charter flight. Captain Maundrell watches from the wing. In the background is the town of Russell.

(Photo via New Zealand Herald)

In the 1960s, when most pilots of his age had retired to the golf course, Charles was not only still flying (now for Pan Am), but had set up his own flying boat airline in the Caribbean. His airline (Antilles Air Boats) operated small Grumman Goose amphibians, but the big Sandringhams at Rose Bay reminded him of old times. He let Stewart Middlemiss know that if ever the two boats went out of service, he would be a prospective buyer.

Several years after Blair's visit, Ron Gillies moved to the Caribbean to work for Antilles Air Boats. Stewart Middlemiss went over for a visit, and was entertained by Charles Blair and his wife, the actress Maureen O'Hara. Middlemiss later described Antilles Air Boats as "a cute little operation, if a bit rough and ready." [2]

So when the end came for the Rose Bay—Lord Howe Island route in 1974, Blair was duly contacted and a sale arranged. *Islander* was to be the first of the two boats to be flown across the Pacific. Then came the unexpected delay through *Beachcomber* going ashore at Lord Howe, and *Islander* being recalled to maintain an emergency service on the route.

This delay was but a minor hiccup, compared to the problems created for Blair by the US Federal Aviation Administration. A senior FAA official from Honolulu flew down to Sydney, and arranged a meeting with the General Manager and the Senior Pilot of Ansett Flying Boat Services. He explained to them that Antilles Air Boats were not renowned for abiding by the FAA's regulations, and therefore they were imposing certain conditions on the transfer of the two flying boats. It seemed that the FAA agreed with the comment made by Stewart Middlemiss about AAB being "a bit rough and ready".

Ansett were told to carry out full maintenance checks on both boats, and issue them with Export Certificates of Airworthiness; also to apply their own usual crew endorsement procedures, before crew members could be issued with a licence for the type. The FAA official would remain in Sydney to see this was all carried out.

Although both boats had been extremely well maintained by Ansett, to issue them with new Certificates of Airworthiness necessitated much work, adding greatly to the final price Antilles Air Boats had to pay for them. In addition to this work, extra fuel tanks had to be fitted for the ferry flight across the Pacific. Stewart Middlemiss, who by now had left Airlines of New South Wales to run Ansett Airlines of Papua

Painted in Antilles Air Boats logo, and renamed Excalibur VIII, the flying boat is readied in the pontoon at Rose Bay for its departure from Australia.

(Photo Lord Howe Island Museum)

New Guinea, later commented: "He paid a modest sum for the aeroplanes, but by the time the Ansett company lumbered him with the total bill, it looked more like he was buying a 747!"

To Lloyd Maundrell fell the task of checking out Captain Blair for the issue of his licence. "During crew training Charles' flying was very sound indeed, he was a fine pilot. However, he failed in his aircraft examinations, and I had to spend considerable time to get him through the second time.

"So the FAA insisted that I fly the aircraft in command to the States, to enable Charles to build up the required flying hours for his endorsement on the type, and I was issued with a United States licence. Charlie was not amused."

All of this work took more than a month. Then, after several days of flying on crew training, and proving of the long range fuel system, *Islander* was ready to say farewell to Sydney for the last time. Her trailing edge fuel tanks, removed after purchase from New Zealand, had been refitted, and a DC3 tank fitted in a wooden cradle on the upper deck.

She was still in the same white, red and grey colour scheme of her latter days with Ansett, but the Antilles Air Boats logo now replaced that of Ansett.

On her fin was painted a black hawk in a white circle. For many years the black hawk had been Charles Blair's personal good luck emblem; now it adorned the fins of his aircraft fleet. The boat carried new US registration letters, N158J, and a new name, *Excalibur VIII*.

The Excalibur name went back to pre-war days. The original *Excalibur* had been a liner of American Export Lines, which was sunk by a U-boat during the war. When the company's subsidiary, American Export Airlines, launched the first of its new Sikorsky flying boats in January 1942, she was given the name *Excalibur*. Since those days Charles Blair had given the name to various aircraft owned or flown by himself, right down to the new flagship of his airline, *Excalibur VIII*.

On the afternoon of 25 September 1974, *Excalibur VIII* took her final leave of Rose Bay. After a long take-off run due to her fuel overload, she turned against the backdrop of the harbour bridge, passed over the base for the last time, then was eastbound over The Heads and into the Pacific.

It was dark by the time she was passing Lord Howe Island, en route for her first landfall of Samoa. This was back into old territory for the boat, although Charles Blair not surprisingly chose Pago Pago in American Samoa as his transit port, rather than the old New Zealand base at Satapuala Bay in Western Samoa. The New Zealand Sunderlands had, of course, finished in the Pacific years back, so flying boat facilities were no longer there.

Pago Pago, with its fjord-like harbour, was reached after 16 hours 10 minutes flying from Sydney. It had once been a port of call for the Pan American Clippers. After fuelling, and a rest 'of sorts' for her crew, *Excalibur VIII* set out for Honolulu on the 27th. With her heavy fuel load, she was able to overfly the old Pan American staging post of Canton Island, on a flight lasting over 17 hours.

Lloyd Maundrell was highly impressed by Charles Blair's navigational abilities during the flight, particularly the way he used his older-type sextant. "For sun sights he knelt on the cockpit floor and took them through the front cockpit window, no mean feat," recalls Maundrell, "He was a magnificent navigator."

That Blair was an expert navigator was not surprising. He had made over 1500 Atlantic crossings — many of them in the flying boat days, when astro was the only available navigation method for the route, and when a crossing could take 25 hours. After the war he had bought himself a Mustang fighter (*Excalibur III*), and with it made several record breaking flights. One of these, from Norway to Alaska, was the first solo flight over the Arctic and the North Pole. Part of the drive which made Blair undertake this flight was that he saw it as the ultimate navigational challenge.

As *Excalibur VIII* neared Honolulu, an incident occurred which has remained vividly in Lloyd Maundrell's memory: "Prior to our arrival in Honolulu, and following a rest in the cabin downstairs, Charles returned to the flight deck. He was now resplendent in his United States Air Force flying gear, with 'Brigadier General C F Blair' across his chest, wings, and squadron emblems. I was on watch in the command seat at the time, and Charlie decided to get into the FO's seat.

"Being a very big man, and with all his gear on, he sat on the four electric airscrew switches, and sent all four engines into full coarse pitch. I managed to extricate Charlie and prevent the aircraft heading for the Pacific Ocean.

"Our arrival at Pearl Harbour was quite splendid. We were escorted onto our mooring by a large United States Navy vessel — a frigate, I think. Trying to taxi in was made difficult by its bow wave ahead of us."

On the 29th they were off again, on the Honolulu to Long Beach sector, which proved to be the longest of the flight at 17 hours 40 mins. There was something of a fuss as they neared the United States mainland, and Los Angeles flight controllers lost touch with them.

The forecast had predicted winds from the west, but they proved to be easterlies — headwinds. To minimise the effect of these winds, they dropped to below 2,000 feet, which put the aircraft off the radar screens and gave them radio contact problems.

However, *Excalibur VIII* arrived at Long Beach with 90 minutes fuel reserves still in the tanks, and the considerable publicity surrounding their arrival soon switched from the FAA's concern over their safety, to the fact that Maureen O'Hara was aboard.

Charles introduced her to the reporters as "our distinguished supernumerary crew member". She admitted to "helping out" in the galley, and added "seventeen hours in that plane is shorter than five hours in a jet". The local newspaper later carried a report that Miss O'Hara looked as though she had just stepped off a Hollywood sound stage.

Blair now had sufficient hours on the Sandringham for Maundrell to be able to sign out his type endorsement, and return to Australia. A week was spent on the mooring at Long Beach, just astern of the *Queen Mary*, to allow for removal of the long-range fuel system and refurbishing of the upper cabin.

By 6 October they were ready to take on fuel for the flight across the United States. Ron Gillies took charge of the operation, and it is still firmly fixed in his memory: "The refuelling operation at Long Beach was not at all funny, being one of the worst I ever had to handle. The local Port Authority would only allow us to fuel at their hazardous materials jetty up the harbour, near the Howard Hughes establishment." (Hughes' great wooden flying boat still sat in its waterfront hangar here, behind a vast locked door.)

Arrival in the United States. The boat passes the Queen Mary as it comes in to land at Long Beach.

(Photo Capt Keith Sissons)

"At the appointed time I arrived off the said jetty, to find a very considerable tide running. So much so that a large vortex had formed off the end of the jetty, right where I had hoped to secure the aircraft. After a lot of hassle and interference from people on the shore, we finally secured the aircraft and took on the necessary fuel."

On 7 October, Ron Gillies and Charles Blair were the pilots as they set out for the last section of the ferry flight to the Virgin Islands. Although Blair was now endorsed for the Sandringhams, most of the captaining of them over the coming years was to be done by Ron Gillies. Because of his age, Blair was not permitted to command a Sandringham with passengers aboard.

Ron continues his story: "We departed for Eagle Mountain Lake, north of Dallas. Here again the fuelling operation tended to get your adrenaline system into action. The tanker was located on high ground, about 400 feet from the aircraft, with a multitude of hoses connected together. In the middle of the operation, in failing light, I realised something was wrong. Our flight engineer, Jim Flanagan, called that he was not getting fuel, but I could hear the tanker still pumping. The hose had disconnected underwater and below the aircraft. I called for pumping to stop.

"By now there were many gallons on the surface around the aircraft. The character operating the fire boat leapt into his boat, surrounded by petrol, and fired up both engines. I headed for the starboard wingtip, expecting an explosion, but fortunately nothing happened — the mixture was probably too rich.

"Next day we headed for Washington, where we landed on the Potomac alongside National Airport. We departed next morning for New York, where we landed in Upper New York Bay, between the Statue of Liberty and the Battery. After half an hour of drifting about, we finally departed for Boston.

"After landing alongside Logan Airport, I taxied up to the northern seawall, being held 6 to 8 feet off the wall by a 10 knot southerly breeze. We refuelled in this position, from a tanker immediately above our bow. There were no problems, and we departed next morning for St Croix, where we arrived after an uneventful eleven hour flight."

Excalibur VIII had given Antilles Air Boats considerable publicity throughout the 9,900 mile ferry flight. However, once the initial excitement of its arrival at St Croix had died down, the event proved something of an anti-climax. The boat was still sitting at its mooring, having not made a single flight, when *Beachcomber* arrived to join her on December 9th.

Beachcomber had also undergone a change of name with the new ownership, and was now *Southern Cross*, registration N158C. With her to the Caribbean had come quite a contingent of Australians; unable to say goodbye to their beloved flying boats, they had uprooted their homes to come with them.

Ron Gillies, of course, had been with AAB for some years now, and was amused by his grandiose American-style title — Vice President Engineering. Bryan Monkton had flown over with *Southern Cross*, and was to stay on as a captain. Noel Hollé and George Alcock had come over as engineering staff to help support the "old boats".

It was ten years now since Charles Blair had established Antilles Air Boats, as a part-time interest while he was still flying for Pan Am. It had grown to be a thriving business, operating about thirteen of the small Grummans on short inter-island hops.

Being amphibians, they could land in the harbours around which most Caribbean towns had grown, lower their wheels, then taxi up a seaplane ramp in a convenient downtown location. Although most of the islands had land airports served by regular airlines, the short inter-island distances and the convenient locations of the seaplane bases meant that it was often the quicker route to take.

From Christiansted on St Croix, the airline served St Thomas (Charlotte Amalie), Tortola (West End and Road Town), St John (Cruz Bay), St Maarten (Phillipsburg) and Puerto Rico (San Juan and Fajardo). They were operating 120 scheduled flights a day, and advertised as "The world's largest seaplane airline". A large proportion of their passengers were island-hopping tourists, but they also carried a considerable number of daily commuters. The busiest route, St Croix to St Thomas, was only a twenty minute hop.

Obviously the big Sandringhams were not suited to an operation such as this. But Charles Blair had grander plans in mind for his new acquisitions. He intended establishing island cruises, with a short flight to a new island each day, between night stops at convenient hotels. The cruises would take in more of the Caribbean than just the Virgin Islands, which AAB currently served. Passengers would be either flown in from mainland United States by jet, or could be collected from an eastern seaboard city directly by the Sandringhams.

Charles Blair also thought that the success of his airline in competing over short distances against landplanes could be repeated on the mainland, particularly between cities such as New York and Boston. The possibility of such services for the Sandringhams was also in his mind, but this was really only an idea — rather than a definite plan — as were the "Down Island Cruises".

Having now safely brought his two Sandringhams to St Croix, Blair soon found out that his problems were only just beginning. He was up against the regulations of the FAA. Before the two boats could be used commercially in the United States, they would need US type certification. There had never been a Sandringham operated in the US. Back in the fifties, South Pacific Air Lines had its Short Solents cleared for operation from California, but a Solent was not the same as a Sandringham.

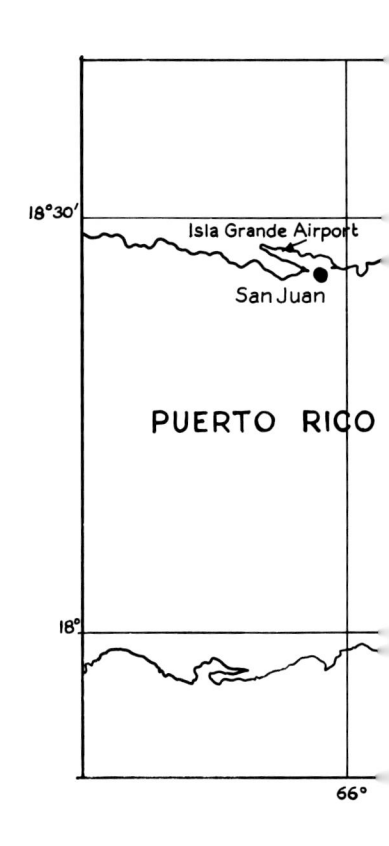

PUERTO RICO AND THE VIRGIN ISLANDS

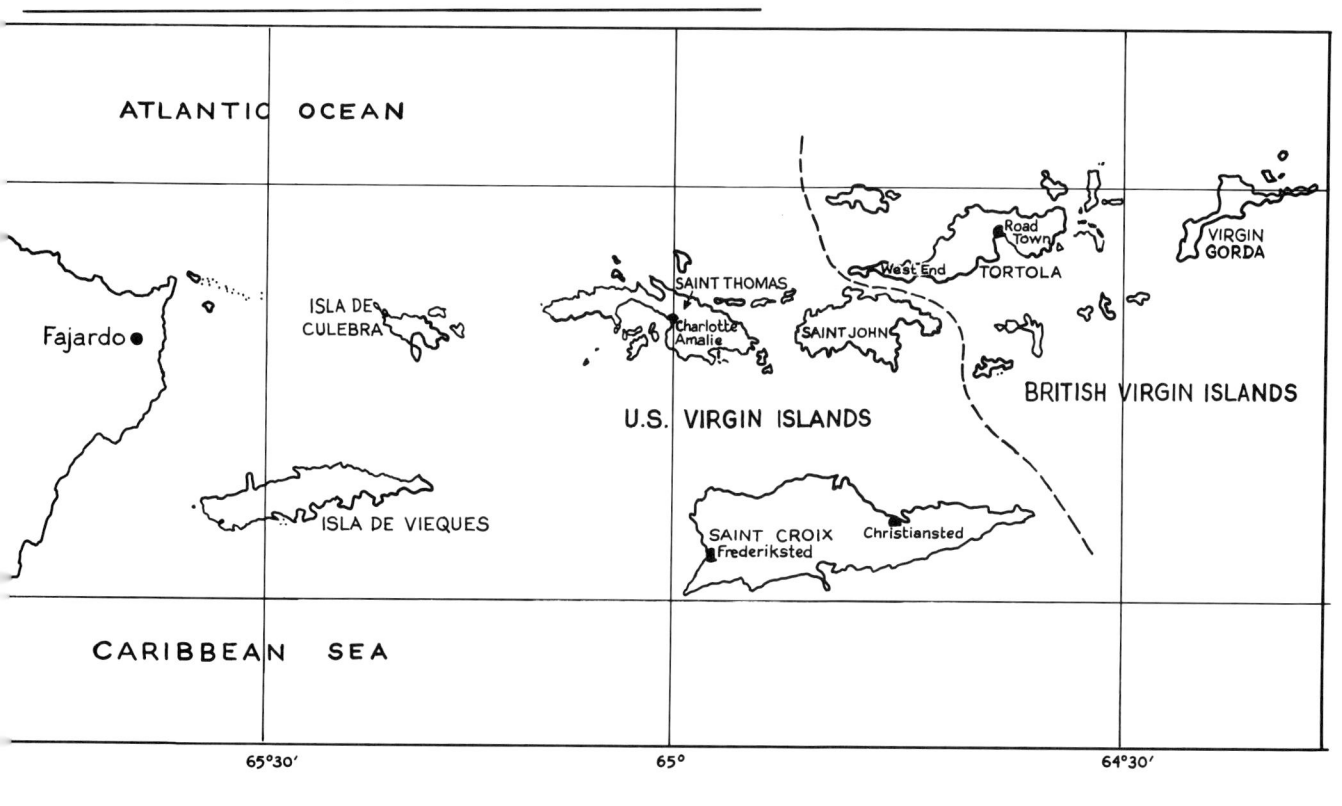

From Christiansted on St Croix, the airline served St Thomas (Charlotte Amalie), Tortola (West End and Road Town), St John (Cruz Bay), St Maarten (Phillipsburg) and Puerto Rico (San Juan and Fajardo).

(Map by Robin Allen)

When Shorts were consulted on the matter, they said that they no longer retained records on the two aircraft, and refused to endorse them for certification. Up against a brick wall with the FAA, Blair looked at the possibilities of putting the two Sandringhams on the British Colonial register. To allow this, a new company, Antilles Air Boats Limited, was established in the British Virgin Islands.

Many hours of negotiations were carried out between John Velox and Les Shelton from the Antigua office of the British Civil Aviation Authority (CAA), and AAB staff. Still Shorts were reluctant to be involved, particularly in the case of *Excalibur VIII*, because of its civil conversion by Ansett and not themselves. Eventually *Southern Cross* was accepted by the CAA, and given the new registration VP-LVE.

But Antilles Air Boats was forced to admit defeat in the case of *Excalibur VIII*. Because it was not a true Sandringham, not even the British were going to accept it. Charles Blair was far from happy. The Australian authorities, who controlled what was by far the safest civil airline industry in the world, had been happy to certify the flying boat. Yet neither the British or American authorities, with safety records well below those of Australia, would do so.

On 15 January 1975, *Excalibur VIII* made what was to be her only flight with Antilles Air Boats following her arrival at St Croix. She was ferried to Isla Grande Airport, San Juan, Puerto Rico. At this old, largely disused US Navy flying boat base, she was put in a hangar, her engines inhibited, and then almost forgotten. Even in trying to operate the one Sandringham, things did not go smoothly for Blair. The British registration did not allow carrying passengers between ports within the US Virgin Islands, and the airline was fined by the FAA for doing so on occasions.

A new home and a new owner. Captain Charles Blair and Excalibur VIII shortly after their arrival at St Croix.

(Photo Fritz Henle)

While operating *Southern Cross*, Captain Blair made two trans-Atlantic flights in her, during the summers of 1976 and '77, accompanied by Captain Ron Gillies. Partly these flights were family visits, as Maureen O'Hara had a residence near Bantry Bay, Ireland. Partly they were nostalgic trips for Charles Blair.

On these Atlantic crossings, Ron Gillies was amused by Charlie Blair's navigation methods, particularly his old sextant which Lloyd Maundrell had commented on during the Pacific crossing. Ron recalls, "The thing was a real antique, and completely unserviceable when he brought it aboard during our Atlantic crossings."

It was from Foynes, an Irish village on the Shannon estuary, that the trans-Atlantic flying boats had set out in the forties. Charles had left from Foynes to make the first ever non-stop passenger flight to New York. Much later, he left from there on the last scheduled trans-Atlantic passenger flight by flying boat. He had been the last to leave and in 1976 he became the first to return.

During these summer visits of 1976 and '77, Charles Blair and Ron Gillies became well-known to English aviation enthusiasts. Both years saw them bringing *Southern Cross* on to England, and operating a series of pleasure flights from Poole and Calshot. People flocked to make a short trip in her, realising that it was probably the last opportunity they would ever have to make a flight in a large passenger flying boat.

While *Southern Cross* continued to do a certain amount of flying, and was even granted limited American certification, *Excalibur VIII* was steadily deteriorating in the hangar at Isla Grande.

During 1978, an English aviation enthusiast, entertaining friends at a conference in Monte Carlo, heard mention of two old British flying boats still in operation over in the Caribbean. Like many others the world over, he was a person who, since

Around the WORLD

Above: A take-off from Rose Bay, against a backdrop of Sydney and the Harbour Bridge.

Below: Islander on its take-off run on the lagoon, Lord Howe Island.

Above: Islander taxiing on Sydney Harbour in November 1972.

Below: Islander at her mooring on Lake Eucumbene in Australia's Snowy Mountains.

December 1974, and both boats are together again. The newly named Southern Cross arrives at St Croix to join Excalibur VIII (foreground).

(Photo Capt Keith Sissons)

boyhood, had always felt a fascination for flying boats. But unlike most others, Edward Hulton was in the fortunate position of being able to own one if he so desired.

Edward Hulton followed the matter up and wrote to Antilles Air Boats, receiving a reply from Charles Blair stating that he might consider selling one of the flying boats. Encouraged by this, Edward Hulton telephoned St Croix on 2 September, only to be informed that Charles Blair had been killed in an accident an hour earlier.

Captain Blair had been taking Flight 941 from St Croix to St Thomas, his fourth trip for the day in Goose N7777V. With ten passengers on board, including a thirteen-year-old in the right hand seat next to the pilot, the Goose's left engine failed five miles south of St Thomas. The aircraft lost height, and Captain Blair attempted to complete the flight in ground effect. However, the left float struck the water, the aircraft cartwheeled, and broke up.

The captain and three of his passengers were killed. One of the United States' most experienced and respected pilots had perished, in an accident which the NTSB report later said should have been survivable. [3]

Despite its twenty or so aircraft, Antilles Air Boats was virtually a family firm; everyone knew everyone, and Charles Blair's death was deeply felt. Edward Hulton decided to let the matter of purchasing a Sandringham rest for a while.

He waited until May 1979, then travelled to Puerto Rico and looked at the two Sandringhams in their rather decrepit hangar. By now *Southern Cross* had also been laid up. He was shown around by George Alcock, the last of the Australian engineers with the company. George, then around seventy, was something of a character. Cousin of Sir John Alcock (of Atlantic crossing fame), George had started as an RAF apprentice, and later qualified as Licensed Aircraft Engineer number 2. After a lifetime in aircraft engineering, he became bored with retirement, and so started working with the flying boats at Rose Bay.

Refurbishment work under way in the hangar at Isla Grande, Puerto Rico. Southern Cross can also be seen in the hangar.

(Photo via E Hulton)

Singlehandedly, George was doing what he could to keep the two Sandringhams in reasonable condition. Although there were other AAB staff at Isla Grande, they all worked in the company's engine overhaul facility which was located there.

Next stop for Edward Hulton was St Croix, where he lunched with Ron Gillies and discussed the purchase of *Southern Cross*. Since Charles Blair's death, Antilles Air Boats had been taken over by the hotel and casino chain, Resorts International, and Ron Gillies has been given the task of disposing of the two Sandringhams.

Ron was amazed that an individual was wanting to buy a large flying boat, and dwelt at length on the certification problems. Then he explained that not even an airline could afford to operate such an aircraft, so how could one person. He soon discovered that Edward Hulton was not going to be easily dissuaded.

Hulton's next move was to bring a seaplane expert (Joe Mangeri) over from Miami, to give him independent advice on the deal. The advice was to be cautious, as although *Southern Cross* looked to be in good repair, there was considerable corrosion present, especially in the planing bottom. Shortly after this, news came through that *Southern Cross* was withdrawn from sale. Apparently a Bahamian millionaire, Happy Miles, was purchasing it.

Worried now that he might lose the opportunity to purchase a flying boat, Hulton looked more closely at *Excalibur VIII*. He made a bid, went to Miami, and concluded a deal with Resorts International's lawyers. *Excalibur VIII* was his. This raised a few eyebrows back in St Croix, as those knowing of the certification problems could not see how anything could ever be achieved with the Australian modified Sunderland.

So was it to be yet another new lease of life for the old World War Two veteran, now thirty-five years old, and assessed to have reached the end of its life three times already? To allow ownership of an aircraft which was on the US register, Edward Hulton established a Miami-based company called Juliet Flying Boats Inc.

By June, Juliet Flying Boats had a small team of sheet metal workers from Miami busy on the boat. Much work was needed. Although in perfect condition when she had left Australia, she had suffered greatly since. The three months on the mooring at St Croix had not been a good start.

Flying boats need to come up the slip, be hosed off inside and out, and the hull checked, at regular intervals. Ansett had aimed at having each boat out once every three weeks. When BOAC had operated them on the international routes, each boat came out at the end of every overseas flight.

Even when safely in the hangar at Isla Grande, *Excalibur VIII* was still partly at the mercy of the elements. There was frequent heavy rain in Puerto Rico, and great torrents of this cascaded into the hangar through various openings in the roof — directly onto the aircraft. It was so bad that those working in the hangar named the two worst leaks 'Victoria Falls' and 'Niagara Falls'. George Alcock had primed areas where there was corrosion, so that she looked like a dirty patchwork quilt, but this was only a partial remedy to a problem which should not have been allowed to develop in the first place.

It was soon apparent that the team working on her were not achieving much. They had removed a lot of hull plating near the rear step, and in the starboard bow area, but that was about it. Ron Gillies visited Isla Grande in the course of his work with AAB, and expressed concern that there was a risk of the hull distorting. By the end of July, Juliet Flying Boats had contracted a Miami-based aircraft maintenance company, Airtech Service Inc., to take on the work. Men, tools, and materials were flown over to Puerto Rico. Under the day-to-day control of the lead man, Doug Sandstrom, the work started to progress on a much more organised basis.

Early 1980 saw the airframe work well advanced, but new problems were coming to light. The four engines, considered to be in good order, all had to go for overhaul. They were supposedly inhibited, yet inspection showed rust in the cylinder bores. They were removed, crated, and shipped to Scottish Aviation at Prestwick.

Ron Gillies had now finished working for AAB, and was employed by Juliet Flying Boats to control the work being done by Airtech. However, Ron now had another interest. The prospective sale of *Southern Cross* had fallen through, and there was a risk of her going for scrap. After spending virtually his whole life working with flying boats, the thought of this was too much for Ron. In a joint venture with some friends from England, he sunk his life savings into purchasing her.

With the assistance of Mike Coghlan, an English garage proprietor who had been involved with the pleasure flights in England during 1976 and '77, Ron was now working to get *Southern Cross* airworthy again. They believed that if they could get her back to England, she would be welcomed there and a role found for her. Indeed a number of volunteers came over from England to help with the work.

So the hangar at Isla Grande was now a hive of activity, with major work underway on the two big boats. Ron was unhappy about some of the work being carried out on *Excalibur VIII*, and believed that his instructions were being ignored. The crunch came when he walked into the hangar and observed a worker dealing with corrosion on the port lower spar boom (the old New Zealand problem) — Ron did not like the way he was going about the work, and so washed his hands of the whole job and concentrated on his own aircraft.

Roll out of the boat ready for launching at Isla Grande, November 1980.

(Photo via E Hulton)

By September, *Excalibur VIII*'s sheet metal work was finished, and the engines had arrived back, ready for fitting. She was looking immaculate in a gleaming new coat of white paint. The only problem now was that a few people connected with the job seemed just a little reluctant to see it finished. Some of them had been there for more than a year now, and had taken to life on the island.

October saw the departure of Ron Gillies and crew in *Southern Cross*, little suspecting the problems and epic flight which lay ahead of them before they would get her safely to Ireland. They were racing to get across the Atlantic before winter closed in and made a crossing by the northern route impossible.

By November it was decided to delay no further, and *Excalibur VIII* was towed out for engine tests and launching — despite the interior being unfinished, and the radios not functioning correctly.

Maureen O'Hara came over from St Croix to crack a bottle of Champagne on the boat in a little pre-launch ceremony. The launch was delayed for a day, however, when one of the main gear tyres burst. When finally launched and some trial taxiing was carried out, all seemed well.

Although *Excalibur VIII* was still on the American register, the FAA were adamant that it would not be allowed to fly in American air space. The only concession they would grant would be to issue a ferry permit, allowing a flight to a foreign territory. So, on 18 November, Captain Don McDermott (an AAB captain) flew the boat out of Puerto Rico to Virgin Gorda in the British Virgin Islands, although with a stopover in St Croix to allow for refuelling at the Antilles Air Boats facility.

For the next four months the flying boat led an itinerant existence around the various isles of the Caribbean and spent some time in the bay at West End, Tortola. Eventually she was ferried back to St Croix and Teague Bay, where Mr Hulton took up residence in The Reef Hotel.

Gradually he was coming to realise that operating a Sunderland as a private air yacht was not a very practical proposition. Firstly there was the problem of finding and retaining a suitable crew to operate it. Then it was necessary to find an airworthiness authority willing to allow it to fly.

A further problem, although no one was aware of it at the time, was that the hull had already started the process of corroding, and if left afloat for too much longer, would end up back in exactly the same state as it was before.

After being proffered advice by many a 'flying boat expert', Juliet Flying Boats came to the conclusion that they would be better off with the boat back in England. There, it was hoped, the airworthiness authorities would be more sympathetic. Besides, with her RAF history, people in general and public bodies in particular would surely be keen to assist the cause of an old 'war hero'.

The first problem was to find a suitable captain for such a flight. The AAB captains were all fair weather pilots. They had neither the IFR ratings nor the navigational skills for an ocean crossing. Bill Mable, a retired AAB captain who had been helping with the boat, suggested that Bryan Monkton (now back in Australia for quite some time), would be the most suitable person to captain her on an Atlantic crossing.

When Captain Monkton arrived in St Croix on 19 February, he had been led to expect that all would be ready for the Atlantic crossing, but this was far from the case. There was electrical and radio work incomplete, and the long range fuel tanks still to be fitted in the upper cabin. His first move was to get rid of a number of 'hangers-on' who had attached themselves to the project.

A month later most of the problems had been solved, and the remainder of the ferry crew arrived. These were Captain Keith Sissons (co-pilot) and Mark Blandford (flight engineer). Keith Sissons was a captain on Short Belfast aircraft with Heavylift Cargo Airlines, while Mark Blandford was a DC6 flight engineer. Captain Monkton takes up the story: "On the 21st March I flew the aircraft out of the little lagoon in Teague Bay for the last time, across to the AAB base at Christiansted, so we could take on fuel for the flight to Bermuda.

"I was still not satisfied with the radios, but had come to the conclusion that no one in St Croix knew enough about them to bring their performance up to a satisfactory standard, and that the only alternative to another lengthy delay was to fly to Bermuda in the hope that they could be properly serviced there."

Because of the time of year, and the fact that *Excalibur VIII* was not equipped with any form of de-icing, Captain Monkton had elected to cross the North Atlantic by the southern route, via Bermuda and the Azores to Lisbon. The biggest problem with this route is that there is no sheltered, all-weather harbour, suitable for flying boats, at the Azores. This meant that they were dependent upon getting reliable advance forecasts of the sea state in the landing area. Back in the days when Pan American flew their Clipper flying boats on this route, they of course had their own station staff at the Azores to provide such information.

On the morning of 27 March, six weeks later than planned, *Excalibur VIII* left St Croix for Bermuda, the Azores, Lisbon, and its final destination, Marseille. On board were the three flight crew, Paul Fagan as bowman and Edward Hulton. It had been intended to take several more passengers, but six was the maximum the FAA would allow under their ferry permit. Bryan Monkton continues: "As a final salute of respect to Maureen O'Hara, we made a low sweep over her house on the hill overlooking the flying boat base at Christiansted, and then set course visually for St Thomas, in order to check the accuracy of our compass."

This was found to be 23 degrees out, so Bryan took an astro compass reading to set course for Bermuda, and adjusted his magnetic compass accordingly. At least it would now read correctly on one heading. "With Sissons at the controls I handled the navigation, and after about three hours took a series of sun sights with the Mk IXB sextant. The resultant fix showed that we were some 25 miles east of track and slightly ahead of flight plan.

The crew gather ready for departure from Teague Bay. Left to right : Captain Keith Sissons, Captain Bryan Monkton, Mark Blandford (flight engineer).

(Photo Capt Keith Sissons)

"Course was altered accordingly and after about two more astro fixes we tracked accurately to Bermuda, alighting in the Great Sound at 1623, six hours thirty minutes out of St Croix. A small boat was waiting for us as we taxied back from the end of our landing run, and the occupants indicated we were to moor up to a very large iron buoy in the middle of the Sound."

The radios had proved virtually useless on this leg, and arrangements were made to have them checked over before continuing with the flight. However, after spending a week working on them, a US Navy technician announced that he could not improve them and that the equipment was obsolete.

To add to the problems, reports coming in from Ponta Delgado in the Azores indicated that the sea state there was not currently suitable for a landing. As both the co-pilot and flight engineer were on limited leave from their regular jobs, they returned to England. Dick Froggatt arrived in Bermuda as replacement flight engineer, he and Bryan Monkton being old friends from Trans Oceanic Airways days. Bryan's reaction was "I could think of no one I would rather have with us." With Dick's experiences while working on P G Taylor's Pacific cruises, he was an expert at away-from-base maintenance.

With a long over-ocean flight ahead, Dick immediately set about giving the essentials of this unknown aircraft a thorough inspection. He was not one to accept at face value the assurance that the aircraft was in perfect condition.

After various other local experts had unsuccessfully tackled the radio problem, Benny Lynch, a radio technician from England, was flown out. He managed to get the vital HF set operational, although it still proved temperamental on the subsequent flight. Various Bermudians gave considerable assistance during the boat's stay there, in particular Wing Commander 'Mo' Ware, the island's former Director General of Civil Aviation.

By now *Excalibur VIII* had been moved from her original mooring in the Great Sound to a more suitable location in the channel between Saltus Island and the shore. A police launch was provided to lead the flying boat through the entrance to this channel.

As work neared completion a test flight was made and, after alighting, *Excalibur VIII* was following the police launch back to her mooring when she ran aground on a coral reef. Luckily she didn't appear to have sustained any structural damage.

Captain Monkton takes up the story again: "While work was proceeding on the aircraft, I had been monitoring the sea conditions in the Azores from the regular reports received by the Met Office at the airport, and it was becoming obvious that the Azores High was going to be late this year, and that we could not rely on the swell falling to acceptable limits for some time.

"On the other hand, Gander was reporting that the alighting area was now free from surface ice, and the freezing level had risen to over 1,000 feet above sea level. So I began to favour the plan to use the northern route through Gander, Shannon, and Southampton. I think this pleased Edward, as he was naturally keen to show the aircraft to his friends in England.

"In some ways I was disappointed to miss the challenge of an open sea alighting in the Azores, but naturally the safety of the aircraft and personnel had to come first. While waiting for work on the aircraft to be completed, I spent some time practising with the sextant at night, in front of the Yacht Club. Although I had done no astro-

Captain Bryan Monkton and Edward Hulton on the flight deck at Bermuda, prior to the trans-Atlantic flight.

(Photo via Capt Bryan Monkton)

navigation for over twenty years, I had little difficulty on the flight from St Croix in taking accurate sights in daylight. But the flight from Gander to Shannon would have to be made at night, and I was somewhat concerned at my ability to recognise the required stars and planets — especially in the Northern Hemisphere — and then to be sure I was looking at the one I wanted through the limited field of the sextant.

"However, after some hours of practice, I found that I could fix the position of the Bermuda Yacht Club with acceptable accuracy, my only difficulty being in finding the dim pole star in the sextant, while still retaining sufficient light from the internal source to see the bubble. Obviously my eyes were not as young as they had been.

"So, on the morning of the 16th May, almost three months after I had arrived in St Croix to make a flight which I had estimated would take a week or so at most, we were ready to depart. At 0830 we slipped the mooring in Saltus Bay and, after taxiing into the Great Sound, were airborne at 0847 and climbing over Hamilton in moderate rain."

Flight plan time for the 1,060 n. miles to Gander was 7 hours 30 minutes, and the co-pilot was now Hugh Watlington, an ex-RAF pilot with post-war Catalina experience. "At 1530 we passed directly over St Pierre Point, and at 1635 alighted on Gander Lake, mooring up to a proper seaplane buoy off a concrete slipway. Canada was obviously flying boat orientated. I had decided that, in view of the disastrously long delays we had experienced on this whole operation, and the great expense to Edward, we should try to press on as soon as possible. Therefore I planned take-off for Ireland the next day, about an hour before sunset, subject to receiving a satisfactory route forecast.

"The following morning, the aircraft was refuelled most efficiently by the crew of a tanker truck from the slipway, the whole operation taking less than an hour as they had obviously done it many times before. At 1705 local time on the 17th May 1981, we slipped our mooring at Gander and at 1720 took off, climbing directly into the setting sun. We levelled off at 2,000 feet, there being a layer of scattered cloud above us. The outside air temperature gauge was reading minus one, and we could not risk the possibility of taking on propeller or airframe ice at this stage by entering cloud.

"As soon as I could hand over the controls to Watlington, I checked our heading with the astro compass, and made the necessary correction to the aircraft compass. Then I took a drift reading, while I could still see the surface of the sea. At 1753 we passed over Cabot Island, which gave me our ground speed, and a wind of SSE at 5 knots. As the flight progressed, the cloud cover diminished, and I was able to get star fixes at frequent intervals. These showed that the wind had veered around to the northwest and that we had a groundspeed of 137 knots. At 0330 GMT we passed from Gander to Shanwick Control. Until then we had been working on Gander HF with moderate success, but we were not able to hear the station in Ireland and did not know if they were receiving our transmissions. We therefore called 'any aircraft' in our vicinity on VHF, and soon had contact with an aircraft above us, who passed our position report to Control. We continued to use this method of reporting for most of the remaining flight.

"At 0400 GMT we were able to climb to 2,500 feet, where the OAT was minus three degrees. It was bitterly cold in the aircraft as, despite Froggatt's efforts, he could not get the cabin heaters to function properly; and this had an adverse effect on our mental and bodily well-being. My biggest concern was the possibility of attracting a build-up of propeller or wing ice, without the means to dispel it. For almost the whole flight from Gander to Shannon, the outside surface of the aircraft was at or below freezing level and, had we entered ice-type cloud or encountered freezing precipitation, we may well have suffered a rapid build-up of ice too heavy or too aerodynamically crippling to carry.

"At 0420, with still some 760 miles to go, I took what turned out to be my last celestial fix as, shortly after, the sky became completely obscured and remained that way for the rest of the flight. All we could do was maintain our present heading, and hope we could get a radio bearing as we neared Ireland. However, despite requests for a QDM from Shanwick Control, they replied 'unable'.

"As soon as it became daylight, I was able to take a drift reading and found this to be 11 degrees to starboard. The wind lanes on the surface of the sea were bearing 320 degrees, and from this I calculated the wind to be WNW at 35 knots. We altered course to allow for this, but, as I had no way of knowing how long we had been in this strong wind, I could only make an approximate correction.

"At 0910 GMT we came over the coast of Ireland at 2,000 feet, and recognised our position as Bull Rock Point Light, showing that we were some 100 miles SW of Shannon and about 30 miles south of track. At 0930 our ADF gave some indication that the beacon at Limerick was dead ahead, and shortly after we came up with this well-known city. At 0955 on 18th May we alighted on Lough Derg. Our total flight time for the 1,810 nautical miles was 13 hours and 35 minutes. So ended what may well be the last crossing of the Atlantic by flying boat."

Despite a successful navigation of the Atlantic, their local navigation was not quite so accurate. When *Excalibur VIII* touched down on what the crew thought was Lough Derg, they were puzzled by the antics of the local people in small craft, who seemed to be waving frantically at them. It turned out that they had landed not on Lough Derg but on the dam lower down the Shannon, and were in danger of hitting submerged obstructions. Once the position was realised a rapid take off was made, although not before some, but not serious, damage had been done to the keel. In the crew's haste to get away, they took off with their flag still flying! Finally they reached the correct alighting area, and moored safely in the river below the lough, near the village of Killaloe.

After several days at Killaloe for a little relaxation and to allow time for fuelling, they took off again for Calshot. Here they were also to spend a few days, with the boat lying at a mooring just off the famous ex-RAF flying boat station, which she had visited so often in her earlier days.

It was during this short stay at Calshot that I first came into contact with *Excalibur VIII*. I had been fascinated by flying boats, and Sunderlands in particular, ever since being given the present of a book on wartime Sunderland exploits when I was about twelve. Being in Australia, I then started to follow the happenings of the Ansett flying boats with interest. I had long since lost track of them when, one day in February 1981, there was an item on the evening television news, concerning the arrival of *Southern Cross* in England. Ron Gillies and his crew, after their epic Atlantic crossing and months of delay in Ireland, had finally made it to Calshot. At the time I was working on a farm in Sussex.

En route from Ireland to Southern France, Excalibur VIII spends several days at Calshot in May 1981. Here in the Solent she taxies past the Royal Fleet Auxiliary vessel Sir Galahad (later to be lost at Bluff Cove in the Falklands War).

(Photo The Times)

Footnotes Chapter 8.

[1] *Red Ball in the Sky* — Charles F Blair (p 109)

[2] Stewart Middlemiss — Interviewed by Greg Banfield Oct '84 *Aviation Heritage* Vol. 25. No. 2

[3] National Transportation Safety Board, Washington, DC — Report No. AAR-79-9

The following Saturday, I tied my kayak on the roof of my van, and set out before dawn for Calshot. And there, to my delight, I saw the unmistakable shape of a Short's flying boat moored offshore. I was one jump ahead of the few other curious spectators gathered on the beach, and was quickly in my kayak and paddling out to *Southern Cross*.

Over the coming months, I was to spend every minute of my weekends at Calshot, helping those responsible for looking after the flying boat. Mike Coghlan, Vic Hodgkinson and Dick Froggatt became my good friends. For the first time in my life I was working on an aeroplane, and a flying boat at that! In those days everyone shared the expectation that *Southern Cross* would continue to fly.

Then in May, Dick Froggatt left for Bermuda to help crew *Excalibur VIII*. When she flew into Calshot, he showed me over her. I watched with envy as the crew boarded her that bright Sunday afternoon, she taxied out into the Solent, and when airborne, made a low pass over the old Calshot hangars before setting course for the Mediterranean.

After a five hour flight, Captain Monkton touched down on the lagoon adjacent to Marignane Airport, the civil airport for Marseille. He had, in a little over three months, completed his task of ferrying *Excalibur VIII* from St Croix to France.

Marignane had been selected as the new home for the Sunderland for two reasons. At the time Edward Hulton lived just down the coast from there, at Monte Carlo. Secondly, and by a lucky co-incidence, Marignane was at that time the only airport in Europe able to handle large flying boats.

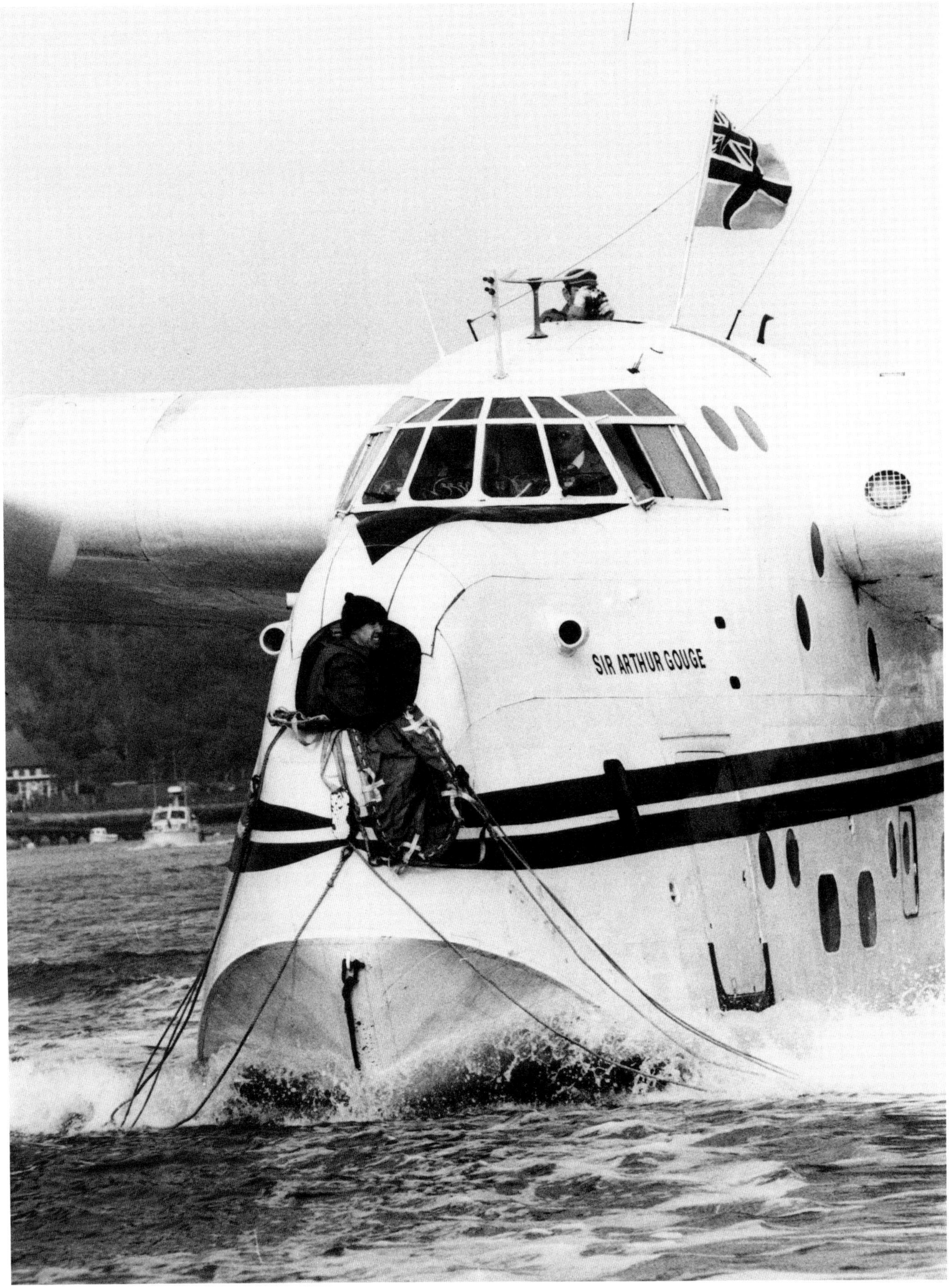
SIR ARTHUR GOUGE

Chapter nine

Back in European Waters

With *Excalibur VIII* safely moored in the flying boat harbour at Marignane, the crew left for home — to various corners of the world. Edward Hulton remained to keep an eye on her, and pump the bilges. This was a very frequent task at the time as, following the several groundings, she was taking in a fair amount of water. In late June, Dick Froggatt returned to organise the beaching of the boat, following the arrival of the shipment of spares and support equipment from Puerto Rico. The beaching was not without problems, due to the very bad state of repair of both the main gear and the tail trolley. Once again the operation was delayed due to a tyre blowing out.

When he had the boat safely ashore, Dick organised repairs on the tail trolley, and for the manufacture of a new main gear wheel. However, while it was on blocks with the tail trolley removed, the Mistral came up, blowing it off the blocks and causing further damage to the keel. Below the waterline, the hull was now in a very sad state of repair. She had, after all, been continually afloat in greatly differing climates for a total of eight months. Just being in the water for this time would have taken a heavy toll on even a well-protected flying boat hull.

But *Excalibur VIII* had long since lost the paint protection put on in Puerto Rico and so corrosion, combined with the physical damage from the groundings, meant that major repair work was now needed. Edward Hulton made arrangements for the work to be carried out by Aérospatiale, in their large hangar on the airport.

Back in England Dick was relating all these problems to me one weekend, as we were working on *Southern Cross* at Calshot. Then he asked: "Are you interested in going down to France to work on her? Edward asked me was I interested, as he feels that he needs someone with the boat full-time. I told him that I would expect a salary of £22,000. After Edward picked himself up off the floor, it was obvious he wasn't going to employ me. So I suggested that you might come a bit cheaper."

Soon I was off to London for a job interview with a director of Juliet Flying Boats. How much was I earning at present? Perhaps naïvely I gave an honest answer, "£72 a week". They would employ me to look after the Sunderland in France for the same. I made the obvious complaint that it would cost me a lot to live there. I was assured that these expenses would be met.

I was employed in the aviation industry! Very marginally, perhaps, as caretaker of an ancient flying boat, but it was exciting stuff to me. The farm job was quickly wrapped-up, then, with my entire worldly possessions (and most of my savings) loaded into my little Fiat van, and my kayak tied on the roof, I set off for the two-day drive to southern France. Not knowing what expenses I would have, or when I would see my first pay packet, I had brought £1,000 in travellers cheques.

The further south I drove, the happier I felt. It might have still been fairly cool, but to come from an English December into bright skies and sun was most encouraging. And when I reached Provence, especially the coastal belt, then this landscape was the nearest thing to Australia I had seen in many years.

Left: Ken Emmott taxies up the Medway towards the dockyard, while the author prepares the drogues. It was a cold day!

(Photo The Times)

The Sunderland and a Canadair CL215 water bomber in the giant Boussiron hangar at Marignane Airport, France.

(Photo Peter Smith)

At the Marignane hotel where I had arranged to meet a director of the company, I found a message saying that he would not be along for two days. The only thing to do was some sightseeing, and learn a little of the locality. Marignane is a small town which had grown largely because of the airport and the aviation industry which developed there. The airport is on the edge of the Berre Lagoon, a vast, relatively shallow expanse of water, connected to the Mediterranean by a short canal.

It wasn't until I visited the modern terminal building at the airport, that I discovered that this area is the real cradle of marine aviation. A local marine engineer, Henri Fabre, made the world's first successful flight from water near here in 1910. A replica of this early, three-floated aircraft hangs in the main hall. Even more surprising, the old gentleman was still alive, and living in Marseille, surely the only surviving member of the original pioneers of powered flight.

Today's terminal, with Airbuses and Jumbos departing for European and worldwide destinations, is half a mile from the original (now semi-derelict) airport buildings, close by the lagoon. Marignane had grown up as a marine airport. In the 1930s, Imperial Airways' Empire Boats came through here on their way to the furthest corners of empire. In the early post-war years, BOAC were here with their Hythes and Solents. Charlie Blair's American Export Airlines flew into here.

And of course the French themselves flew from here. Pioneering flights had set out from Marignane, notably Jean Mermoz in his famous Latécoère *Croix du Sud*. Following the pioneers came the Aéropostale and other civil operators. Marignane was the natural departure port for all Mediterranean flights, and particularly the important route to Algiers.

Flying boat manufacture had also developed at Marignane with the firm of Lioré-et-Olivier (later Sud Est). After the war Vampires were licence-built here, and then part of the Sud Aviation Caravelle. Now the factory was Aérospatiale's helicopter-building division.

But perhaps the airport's most surprising feature is that it is still geared up to flying boat operations. Two tenders are kept afloat, manned by the airport fire crews. The slipway is as clean and well-maintained as it would have been in 1938, and the moorings are still ready for use in the flying boat harbour. Because the lagoon becomes as rough as the open sea when the Mistral is blowing, an artificial flying boat harbour had been constructed.

The reason for all this is that Sécurité Civile, the French civil disaster organisation, base their Canadair water bombers here. Although the Canadair 215s are amphibians, and normally use the land airport, the water facilities are maintained in case of emergency. The Canadairs could be seen daily, practising water scooping and dropping on the lagoon.

Eventually the director arrived and we went to see the Sunderland. It was in Aérospatiale's massive concrete hangar known as the 'Boussiron'. The size of this hangar, heavily built to withstand the Mistral, was such as to make the Sunderland look tiny, as she sat in one corner surrounded by bright yellow Canadairs. The hangar had originally been built for the construction of giant flying boats, in the same era as Britain was building the Saunders-Roe Princess.

Next we went to the opposite side of the airport, behind the modern Aviation Générale hangar, and by the lagoon. Here, in an open-fronted shed, were all the Sunderland's spares, strewn all over the floor where they had been pulled out of their packing cases by local customs officials.

I looked at them in disbelief. Here was almost the entire world's supply of Sunderland spare parts, from flap motors and propeller blades to flight panel instruments, all of which had passed to Juliet Flying Boats by the previous owners of the boat. Many of the components could not be bought anywhere, would be essential to keep the boat flying, and would cost a small fortune to have manufactured today. Yet here they were, open to pilfering by anyone on the airport, not to mention what the weather would soon do to them.

Having been shown around, I asked what I was expected to do. I had come down here with no real idea of what work was expected of me, although Dick Froggatt had suggested several cleaning tasks that were needed on the boat. I didn't really have the experience to do much more. But it was fairly obvious that these spares needed sorting. My new boss was only going to be around for another day, then I wouldn't see him for at least a month.

Certainly we needed some secure area as a store and office. Not only were the spares spread everywhere, so also was the aircraft's vital documentation. Eventually it was decided to rent some space and we came to an arrangement to share half of a secure store/workshop area with the airport's radio engineer.

Soon I was left to my own devices. I shared my time between the aircraft and the spares. I scrounged some old Algerian packing cases from a freight agent and turned these into shelves for the smaller components, then started moving the spares in my van. The packing cases provided office shelves and a desk (which I am still using today), and gradually I got the paperwork into some form of order.

Across in the Boussiron, I was less certain of how to proceed. I was very aware that the hangar was full of highly-skilled French aircraft workers — busy on the Canadair maintenance work — and probably there would be curious eyes watching what I did. The sheet metal workers who had been replacing the twenty feet or so of damaged keel were now close to finishing this work. They had produced their own large and detailed drawings of the area of the hull they were working on. Although no doubt we had original Shorts' drawings covering this, amongst the mass of paperwork, there was little hope of finding them.

The Aérospatiale men showed me other areas of bad corrosion which they had marked on the drawings, then took me down into the bilges to point the problems out. They considered that there was much more work which should be done. It was only too obvious that the long period afloat had taken an enormous toll on the condition of the hull plating. Soon their work was wrapped up and the little team departed for Paris. They had explained that although they also had to paint below the waterline, the hangar here was unheated, and it was too cold for several months yet to allow spraying.

I made a start on one of the jobs Dick had suggested. Borrowing a stand from the French workers, I started to clean the engines and then brush on a metal preservative. The newly overhauled engines had been fitted to the wing just as they arrived from the overhauler. This is all very well on a landplane, but on a flying boat, where they will be continually doused by salt spray, a heavy layer of protection is needed. The relatively new engines were now in a sorry state. Steel fittings were rusted, alloy components badly corroded. The light alloy cooling baffles had corroded right through in many places.

Despite my own French being limited to a few basic words, Aérospatiale's hangar foreman made himself known to me shortly after my arrival, and we struck up something of a conversation. It turned out that his first work, many years back when he joined the French Navy, had been on Sunderlands.

On showing him through the boat, I discovered that the Paris team who had been working on it were under instructions not to allow any of the 'locals' aboard. Then I learnt that most of the older men working in the hangar had served their time in the Aéronavale, on Sunderlands. Soon I had a steady stream of keen visitors. One in particular, an old Breton, would waylay me almost daily, to reminisce on the magnificent times he had enjoyed in Sunderlands.

M. Sliwinsky, the foreman, became a regular visitor, often spending several hours a day with me on the Sunderland. There seemed little point bluffing, so I confessed to having never worked on aircraft before. This was the invitation he wanted. He suggested tasks that needed attention, in particular finding areas of corrosion and teaching me how to deal with them.

Gradually I came to realise how lucky I was. The Sunderland was in the only hangar in the Western world where a big fleet of large marine aircraft were still being maintained, and to absolutely first class standards. Not only was there a

Around the WORLD

Above: G-BJHS moored on the Thames against a backdrop of the Tower of London, September 1982.

Below: Preparing to leave the Thames mooring for Calshot, October 1982.

Above: Departure from the Thames for Calshot, October 1982.

Below: G-BJHS flying up the River Medway after the move from Calshot.

wealth of continuous flying boat experience here (going back 70 years), there was an unrivalled knowledge of the application of modern materials, chemicals, and lubricants to marine aircraft.

So I set out to learn all I could from those around me, and in doing so was quickly accepted into their ranks. This was perhaps aided by my always taking care to point out that I was Australian, not English. I looked forward to lunch in the 'works canteen', if it could be called that, as the food was like that of a first class restaurant. Here I could listen to their conversation, which often came around to the 'Sunderland days' for my benefit.

My friendship with M. Sliwinsky grew, and I was invited to a number of meals with the family. His wife worked with Bureau Véritas, the French airworthiness authority. Later, after the Sunderland returned to England, he was to visit me there. From him I learnt a lot about French aviation history, and of the Marseille area in particular.

When the director paid his first return visit he did not have a lot to say, although he did seem surprised at the 'splendour' of my packing case office furniture. Still, at least he did not express any dissatisfaction with what I had been doing.

One important task was the settling of the accounts, which was to become a 'routine' ceremony. Firstly, I would produce my petty cash receipts for any expenditure, which was never very much. This would be added to the wages owing and an amount would then be deducted for other items on the hotel bill, including my breakfast and evening meals.

My accommodation expenses covered bed only, and by the time hotel food bills had been deducted, my pay was pretty meagre. In fact, when we arrived back in England and I balanced my cash against remaining travellers cheques, I was not surprised to discover that it had actually cost me money to work in France for eight months. Still, I had had the use of a flat in Monte Carlo for a weekend, along with a free ticket to the Grand Prix, so I could hardly complain.

By early 1982, it had been decided that the best way ahead was to try to obtain a British Certificate of Airworthiness, and preliminary discussions with the Civil Aviation Authority had been encouraging. Although this could be done with the aircraft based in France, it made sense financially to move it to the UK. To allow this, an advertisement was placed in *Flight International* for English-based flight crew and maintenance engineers for the Sunderland.

Another company, registered in Guernsey and known as Sunderland Limited, was formed. Control of the aircraft passed from the Miami-registered Juliet Flying Boats to Sunderland Limited. This was necessary legally, to allow the aircraft to be put on the British register. I removed the old US letters, and put on her new registration G-BJHS.

February saw the arrival in Marignane of the company's new Chief Engineer, Bill Mares. Bill was a flying boat man through and through, who had spent his entire life in aviation. Starting as a lad with a flying circus in the 1930s, he later became a ground engineer with Imperial Airways. Then, with the arrival of the flying boats, he transferred to flight engineer. He still held, and was proud of, his flight engineers licence number 18.

Of course I was sad that my mate Dick Froggatt was not more directly involved, as I considered that he probably knew more about flying boat maintenance than anyone. However, the company had no choice but to turn to someone else if we were to obtain a Certificate of Airworthiness. Dick did not hold the necessary licences.

Bill Mares had started in British aviation in the days when there was a saying "To get ahead in aviation wear a Trilby", and he still believed in the maxim. Even when the Mediterranean summer arrived, Bill was to be seen clambering over the Sunderland in collar and tie, cardigan, and Trilby, to the undoubted amusement of the French. It wasn't long before they nicknamed him Le Chapeau (Dick Froggatt, who sported an RAF-type moustache, had been the obvious La Moustache). No doubt I also had a nickname, but never discovered it.

Over a number of visits and several months, Bill checked through the essential systems of the aircraft in preparation for the flight to England. It would be his responsibility to sign out the aircraft as fit for the flight, and we were still making new discoveries about previous work.

Easter saw the weather warm enough for Aérospatiale to carry out the painting, after which the boat was moved out of the hangar. Bill was now able to run the engines for the first time, and after a few minor adjustments all seemed well in that direction.

By late-June things looked ready for the flight to England. All the arrangements were in place, and for a start we would land on the Thames and spend a while moored in the heart of London. The Falklands conflict was just over, and a generous offer had been made to show visitors over the boat to help raise funds for the South Atlantic Fund (for war victims and their families).

We launched the boat with the aid of the airport Sapeurs-Pompiers (firemen), and one of their launches. Once it was on the mooring, there was the beaching gear to remove. The only other person available to help me was Andrew Evans, one of the new pilots. With only a six-foot plastic boat and a 1·5 hp outboard to tow the heavy legs ashore we seemed to manage the job, although looking back, I wonder how. At least it was calm on the lagoon, but I would never tackle the job so light-handed today.

Once the boat was afloat, I moved out of the hotel and started sleeping aboard, although I still returned for evening meals. There was no way that I was going to miss the excellent cuisine at the little Hotel l'Oasis, which had been home for the last seven months.

A flying boat at its mooring always lies head into wind, so there is no airflow aboard. Thus the cabin temperature was usually pretty intolerable, so I took my sleeping bag up onto the wing and slept there. This I found most enjoyable, although I was woken most nights about 1.30 am, as the evening postal flight took off and passed just overhead.

Bill came down to take charge of refuelling. It seemed a bit of a performance at the time, but looking back it was by far the easiest refuelling we have ever done with the boat. Selecting a windless day, we simply hand-warped the boat from buoy to buoy down the harbour until she was tail on to the jetty. An airport tanker drove onto the jetty, its hose was passed across, and the fuelling completed in several hours.

Now we were ready for a test flight. On the allotted day, and with the two newly-recruited pilots up front, we set out from the mooring. Taxiing from the harbour into the open lagoon, the captain brought her into wind and gradually opened the throttles. She gathered speed, lifted onto the step and then, at around 60 knots, started to porpoise badly. It was somewhat unnerving, to say the least, until he throttled back. A second and third run were made, with the same results. We returned to the mooring.

Something was drastically wrong, but no one knew what. Was it the aircraft? Was it the pilots? They were both current airline pilots, both with a wide experience on flying boats, although their boat experience had been some years back.

With the boat now having been afloat for well over a month, quite a growth of barnacles had developed on the hull. I thought that this may have contributed to the problem, and pointed it out to Bill. "No," he replied, "Don't worry about that. I've seen them twice as bad."

It was decided that the only solution was to get a more experienced pilot. I shall let Ron Gillies continue the story: "I was busy helping a friend, whose Caribbean commuter airline was having a problem or two, when I received a phone call from France. Would I be prepared to ferry their aircraft from Marignane to the UK? I agreed, and several weeks later my ticket and more details arrived.

"On arrival in London I met Edward Hulton, with whom I had worked several years earlier, and we proceeded to the Port of London Authority HQ, where we met the gentleman handling our operation. Edward had already done all the groundwork, so my job was a lot easier than I had expected. With all the details of our arrival settled, we departed for Marseille, where we met up with Peter Smith, who had been maintaining the aircraft."

What Ron omitted to say was that when he was in London he went up and down the river by boat, and then walked the banks, until he knew every bit of the river that he was likely to be manoeuvring on. He studied it at every state of tide he expected to meet, until he probably knew it as well as some of the rivermen who had spent a lifetime on it.

On his first run out to the Sunderland, intended to be a test flight, we went nowhere. Ron took one look at the boat and said, "It is loaded wrongly." Although at that time we had no load sheets available, one glance at how the boat sat in the water was enough to tell him that the centre of gravity was too far aft. He came aboard, looked at the considerable quantity of spares I had stowed, and told me what to move and where to stow it.

That sorted, we were ready for another try. Ron continues: "After a satisfactory run-up, I lined up for take-off, and on receiving a clearance from Marignane Tower, applied the power and we were away — or so I thought. We accelerated very slowly, considering our light weight, and as the speed increased I had to apply back stick, until the control column was hard back on the stops, just to run level. We reached 75 knots and stabilised at that speed.

"Obviously something was very wrong, so I gently eased the power off and taxied back to the mooring. I knew that the loading was now OK, but I had another idea. I moved smartly down to the front hatch and, leaning overboard, ran my hand along the bottom as far as I could reach. Sure enough there were small mountains of calcium on the planing surface. What hope of reaching take-off speed with all that growth on the bottom."

Ron Gillies taxies under the raised ramps of Tower Bridge and starts his turn to put the boat between the two mooring buoys. The port drogue is deployed to help the turn, the starboard drogue hangs ready on the bow. Bill Mares surveys the scene from on top.

(Photo Gary Weaser / The Guardian)

A team of divers, drawn from the airport firemen's emergency crew, were organised to scrape off the bottom. I cut them some wooden scrapers from plywood and, armed with these, they managed to get the bottom quite clean. Departure was organised for the following morning. This was Friday, 6 August, the day that had been set for our arrival in London. There would be no test flight now, it was a case of straight up and away, we hoped.

Ron Gillies continues: "The day dawned bright and sunny in our area, but it was very poor along my intended route, direct Marseille to Cherbourg. I therefore elected to track via Narbonne, Toulouse and then visually along the coast to St Nazaire, before heading for Nantes, Rennes, and Cherbourg. Before departing, we were honoured by a visit from the Mayor of Marseille and his entourage, to say goodbye and wish us luck. Our crew were myself, Andrew Evans as first officer, and two flight engineers, Dick Froggatt and Bill Mares. Edward Hulton and Peter Smith made up the complement.

"This time we got ourselves airborne without any problems and headed for Narbonne. The weather in this area was beautiful, and we enjoyed having a reasonably close look at the countryside, through to a short distance north of Bordeaux. From there on the weather began to deteriorate quite rapidly. I had wanted to see the harbours which had been U-boat bases during World War II, but this was not to be. I eventually abandoned that idea and climbed to the LSA and resumed our IFR track, via Nantes, Rennes, and Cherbourg. By the time we were over the Channel, headed for Seaford, we were sitting on top of a layer of strato-cumulus at 3,000 feet, once again in the sunshine. From on the top Seaford we headed directly to Biggin Hill. Quite suddenly, the cloud disappeared and we were over the south-eastern suburbs of London. As we circled the centre of the city we were joined by an RAF Nimrod from 201 Squadron (ML814's old squadron) and a Royal Navy Wessex helicopter. It is not very often that you can indulge in a bit of low flying over any city, and I was enjoying this small benefit of the trip.

On the mooring, directly opposite the Tower of London.

(Photo François Prins)

"We were now in radio contact with the PLA launch, which stopped all river traffic in preparation for our landing. We had arranged two possible landing areas. The Lower Pool, from Cuckolds Point to Wapping Pier, was favourite. But, because an easterly wind of 12/15 knots had suddenly arrived in the area, we were now to use the eastern runway, by Shadwell Basin. This runway was very short owing to the height of buildings across each end, together with the lack of any escape route — as the downstream end was blocked off by a large number of pleasure boats, of all shapes and sizes. With a very light aircraft, I began final approach at 90 knots, with full flap extended, leaving two rather tall apartment blocks to port. This still left us with a very poor approach path, as the buildings at the water's edge were considerably higher than I would have liked.

"As we arrived at a position about five feet above the water, I realised that the wind had switched back to the north at about 15 knots, and we had taken on considerable drift. I tried to kick this off, but was too late. The result was the worst water landing I have made in the last 50 years. It was so bad I felt compelled to apologise to the crew. Without further incident we turned and taxied upstream to Tower Bridge. It opened as we approached, and presented quite an impressive sight as we passed under the raised ramps. Our mooring was immediately downstream of the cruiser HMS *Belfast*, opposite the water entrance to the Tower of London. It was only after we had moored, that I realised that we had an audience of thousands of people lining the river banks for some considerable distance."

From my viewpoint as a passenger, the arrival had been quite memorable. The apartment block Ron mentions was, I think, offices, as I clearly remember seeing rows of people lining the windows and staring in disbelief as we flew past and they could look down on us. The old dock basin we passed over just before touchdown was in use as a watersports area, and it was school holiday time. I vividly recall the startled faces of children in canoes and sailing dinghies, staring up as we rumbled directly overhead at not more than a couple of hundred feet.

After we had alighted, I went forward ready to deploy the drogues and pick up the mooring. From the bow hatch I could see that Bill Mares had opened the astrodome and was standing on top (wearing, of course, *the* Trilby). Once through Tower Bridge, Ron had to avoid a large ship's mooring buoy, then swing the Sunderland hard to port and place her between the two buoys which were to be our mooring. Much to my relief, he carried this out perfectly, and we were soon safely moored.

There were crowds of newspaper and television reporters to get on board, and interviews for Ron, but eventually things quietened down. The crew left, and I was able to survey the new site of my home for a few months to come. It was certainly a very central location. I thought it wiser to give up sleeping on the wing.

Next day, Ron returned to be interviewed by some aviation journalists. As he looked through the newspapers, all of which carried photos of our arrival, he shook his head. "I have waited a lifetime for the opportunity to land a flying boat on the Thames in London, and every single photo is ruined by that man on top in his stupid hat." Ron had now landed ML814 in both central New York and central London. I would think it unlikely that any other pilot has achieved this unusual 'double'.

Of course the visitors arrived, in droves, and milled about on the river banks. They had come to see over the Sunderland, as the papers had said they could. But whereas I thought all had been arranged for the London visit, the only thing that in fact had been arranged was the landing. No one had even considered that the visitors would need to get from the shore to the Sunderland.

The local pleasure boat operators had expected to make a killing, ferrying visitors out. But all their craft were too big for the job. All we had was the plastic Little Blue Boat, and it was overloaded with two aboard. My kayak was proving ideally suited to a flying boat, but it was only personal transport. I used it to go across to HMS *Belfast* for a shower, and usually got ashore at least once a day for a decent meal at a local cafe. On the river I could leave the kayak moored alongside, but in rougher conditions it lifted in easily through the starboard hatch and stowed in the tail.

After two weeks another boat was procured. It was not really the ideal craft to be a workboat for the Sunderland, and I was disappointed that I had not been consulted on the purchase. It quickly became *The White Boat*, and was the only craft we ever had for tending the Sunderland. It has proved sturdy and reliable, but is far from adequate for our needs.

Now, at least, we would be able to get all those visitors across. By now several members of the Aircrew Association had become regular visitors to the Sunderland, and they were willing to help with the ferrying and showing around of visitors. But the pleasure boat operators wouldn't allow us to use their landing stages. They were annoyed at not getting the trade themselves, despite having no craft suitable for coming alongside the Sunderland.

Not to be beaten, we started collecting visitors from a steel ladder set in the side of a nearby wharf. It was a long, vertical descent at low tide, limiting our visitors to the brave and physically able, but we could get some over. After a day, the Board of Trade moved in to stop us. The pleasure boat operators had lodged a complaint.

Because we made a charge to see through the Sunderland, that was interpreted as a charge for the boat trip across, and our boat wasn't licensed. All right, we would licence it. No that was not possible — it was below the minimum size to be licenced. In the end we continued to take people across, but without any charge. There was a box for donations. So much for the grand plans to throw the Sunderland open and make a fortune for the South Atlantic Fund. But we did take a little for Service charities.

The episode did not even end there. Except for one very helpful boatman who provided a standby tug for our arrival and departure, we suffered continual harassment from some of the Thames rivermen. Pleasure boats moving up or down the river would pass within feet of our wingtip. On occasions they brought their boats across the river and left them against the ladders used by our visitors, preventing our access. All-night disco boats would moor as close as possible to the Sunderland.

But the boatmen weren't the only hazard on the London River. A lot of very heavy debris came down with the tide, the worst of it being heavy baulks of timber from dilapidated wharves. Most would be carried around us, but the odd piece hit the hull. It would have only taken an unlucky hit by one of the heavier pieces, some of which had massive steel spikes protruding from them, to badly hole the Sunderland.

Also, I was far from happy with our fore and aft mooring, which the PLA had insisted we use, despite my objections. The arrangement put excessive stresses on the Sunderland's hull and mooring fittings, especially at low water when there was a lot of slack in the riser chains, or in a strong cross wind. Not surprisingly then, I was getting anxious to see the boat leave the Thames.

Gradually, however, I came to realise that no plans had been made for it to go anywhere. Only now were the possibilities of getting it ashore somewhere down river in dockland being looked at, but no suitable sites were found. To me, the obvious thing seemed to be to take it to Calshot, which was both suitable and available. For some reason which I could never determine, there was a very deep resistance to taking it there.

Finally a suitable location was found for the boat. Shorts had offered outside storage at Belfast during the coming winter. She was to return to her makers. There was just one minor problem. We needed to refuel for the flight, and the PLA wouldn't allow it on the Thames. Fuelling an aircraft should be no more dangerous than filling your car, probably less so, as those doing it are normally trained for the job. But we couldn't change the minds of the PLA. So a refuelling stop at Calshot was arranged en route for Northern Ireland.

Ron Gillies returned from his work in the Caribbean, and at 4 pm (to be at the best state of tide) on 19 October, we left the mooring and taxied down river for take-off. After a short flight to the South Coast, we alighted in a fairly rough Solent.

Two Belfast-built vessels moored together on the Thames. The Sunderland astern of the cruiser HMS Belfast.

(Photo François Prins)

As Ron taxied in past the *Calshot Spit* lightship, I opened the bow hatch to prepare the drogues, only to get drenched to the skin by the breaking seas. At least conditions were not so bad once we were in Southampton Water, and we picked up the mooring safely. Waiting for us in the Solent, and escorting us in to the mooring, had been Mr Etheridge in his boat *Prince II*. At Calshot, we were lucky to have one of the few boatmen still around used to handling a flying boat. Mr Etheridge had tended *Southern Cross* when Captain Blair had brought her here in '76 and '77. Once we were moored, Mr Etheridge ran the crew ashore, leaving me to 'batten down'.

There were various covers to fit, the mooring light to rig, and bilges to check. The vital storm pendant had to be shackled and moused, between the bow and the mooring buoy. Then, finally, all hatches to be double-checked. It wasn't that the rest of the crew wanted to abandon me to these jobs — they had little choice. Whenever we arrived at a mooring, there would be a safety boat standing by, and he would be anxious to ferry everyone ashore and then get away. I had my own transport (the kayak), and so would be left to it.

It was well after dark by the time I made the wet paddle ashore, then walked the one-and-a-half miles up to what had in RAF days been Top Camp, and was now Calshot village. Here, I joined the rest of the crew in The Flying Boat pub, once RAF Calshot's officers' mess. Inside, it could have still been an RAF mess. On the walls hung photos of early flying boats, the Schneider Trophy seaplanes, and of course officers of the RAF High Speed Flight, which was based here between the wars. I loved Calshot, with its aviation memories everywhere, and wished we were staying here instead of leaving for Belfast.

Next day we taxied up Southampton Water to Hythe, where Mr Etheridge had arranged our refuelling. We were towed in to a jetty where the barrels of fuel were waiting, and used an electric pump borrowed from Flight Refuelling Ltd. Halfway through, the batteries ran flat and the local vicar, who had dropped by to watch operations, borrowed some more to get us out of trouble.

Take-off from the Thames, bound for Calshot.

(Photo François Prins)

For me, the best news of the day was a telegram from Belfast. Shorts announced that they could no longer accommodate the Sunderland for the winter, no reasons given. Surely we would now bring her ashore at Calshot. But no, the company now conducted a survey of potential flying boat bases in Europe. Sunderland Ltd had gone to great effort to bring the Sunderland back to England, in order to get a British Certificate of Airworthiness. Now it looked as though it would go overseas again.

Ron Gillies arrived back, weeks later, having apparently toured every old flying boat base in Spain, Portugal, and Italy, but without success. He had almost been shipwrecked when he was caught in a severe storm on a ferry in the Mediterranean. Ron, not normally one to complain, said it was the most frightening sea voyage of his life. Finally, the decision was made to bring the boat ashore at Calshot.

But firstly, the beaching gear had to be brought over from France. By the time all was in place for the beaching operation we had been on the mooring at Calshot for almost two months. On 14 December, the army generously provided a heavy winch vehicle from 17 Port Regiment, Marchwood Military Port, and the Sunderland was hauled up the Calshot slip once more. It had been 37 years since she last came ashore here.

I wasn't sad at seeing my home on dry land. Its hull was in a terrible state after six months afloat. The new coat of protective paint put on by Aérospatiale had flaked off in sheets soon after launching. The interior was now running with moisture, the cabin upholstery like wet sponges. And what a relief not to have to make those freezing wet trips back and forth by kayak.

Although many of the old RAF hangars were still standing at Calshot, they were all utilised by the Activities Centre which now occupied the site. Some were used for

Taxiing in from the Solent to the Calshot mooring, October 1982.

(Photo Southern Evening Echo*)*

storage of sailing dinghies and canoes, one even housed an indoor cycle racing track and a ski slope. So the flying boat, and those working on it, would have to stay out in the elements.

There hadn't been much I had been able to do in those two months on the Calshot mooring. The APU had to be run regularly to keep the batteries up, and the engines turned. Bilges, floats, and moorings to check, and that was about it. I had been eating in the excellent canteen at the Activities Centre. Evenings I had usually spent in their bar, and then would set out for the dark, and often rough, half-mile paddle out to the Sunderland. Surprisingly, the only mishap in that time involved my typewriter. I had brought it ashore one day to write some letters. On the return journey, when making the awkward transfer from canoe to flying boat, the canoe capsized. It was goodbye typewriter.

Now we could get stuck into the task ahead, getting a Certificate of Airworthiness. I located two caravans for the company to purchase, one for £40 to turn into a workshop and one for £300 to become my home. Our local CAA surveyor, Mike Inskip, came along to discuss the way forward. And were we lucky — he was the last man in the CAA who had once been a flying boat engineer.

It had been decided to apply, in the first instance, only for a Private Category C of A. There seemed no major problems in obtaining this. The aircraft would of course have to be 100% sound, and some systems would have to be modified to comply with current regulations. To obtain a Public Transport C of A, however, would be a different matter. Not only would all sorts of extremely expensive modifications have to be made, to meet current airliner standards, but an extensive structural survey would also have to be completed. Even to get the Private Category C of A there was much work ahead. Hardly surprisingly, Mike Inskip didn't like the look of the planing bottom. We would need to arrange a survey of the hull plating. Over the years many modifications had been carried out. Some of this work had been completed without supervision by an airworthiness authority. Did these changes comply with regulations, were there drawings, had approved components been used?

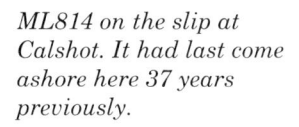

ML814 on the slip at Calshot. It had last come ashore here 37 years previously.

(Photo François Prins)

The biggest headache would be the electrical systems. We knew that since leaving Australia, all of the Sunderland's eleven manual fuel cocks had been replaced with electrical ones. But the flight engineer's indicator lamps were connected to the selector switches. They merely told him if he had selected open or shut — the valve itself could be to any position. Much more wiring would have to go in, microswitches connected, the valves modified. The fuel cocks are only an example. There were many other 'modifications' to make good.

But even more fundamental electrical changes were needed. Modern systems philosophy was very different from that accepted when Ansetts had converted and rewired the aircraft in 1964. All the aircraft's electrical systems were connected to a single bus, or supply. Should a fault develop in a heavy current-drawing component, or perhaps a cable go down to earth, it would take out the entire electrics. We would have to break up the system with separate essential and non-essential buses, an emergency bus, a radio bus, an emergency radio bus, and so forth.

After an afternoon discussing all this by the wood-burning stove in my caravan, Mike Inskip was asked "How long would you expect all this work to take?" "Ah", he replied, "that depends on how many Indians you have on the job". Indians, I thought to myself, why will we need anyone else? With me working full-time, and some part-time help by specialists on jobs like the electrics, we will soon get it done. Little did I know what lay ahead...

Work got underway. Bill Mares made regular visits to supervise things. Jon Street, a member of the Calshot Lifeboat Crew, became a regular part-time helper. The propellers came off and went for overhaul. Ron Albutt, an aircraft electrician who had recently taken retirement, joined us to tackle the electrical problems. Dan-Air Engineering, from Lasham, sent down their NDT (non-destructive testing) technician to survey the hull.

To our great relief, this showed all the hull plates to be within one or two thousandths of an inch of their original thickness. But the hull would be an ongoing problem. Normally the hulls of civil flying boats were not painted, relying on their anodised finish for corrosion protection. But twice now the boat had been left afloat for excessive periods of time, without any form of protection once the paint came off.

Jon Street cleaning off the heavy layer of marine growth.

(Photo François Prins)

Not only had most of the anodising been lost, but much of the aluminium cladding had been removed from the dural plating. This left the hull extremely vulnerable to further corrosion, particularly below the waterline.

Two sheet metal workers from Dan-Air started working at weekends, to replace badly corroded hull plates. But a solution had to be found to the problem of effective protection. I investigated finishes used on aluminium ships, and decided to experimentally apply a marine-type finish to the forward section of the planing bottom. On the rest, we reverted to the traditional flying boat coating — beeswax and lanolin.

One misty day in March 1983, an unusual craft went past ML814 which, being right on the end of Calshot Spit, was very obvious from passing vessels. This craft was a large army pontoon, on which sat ML814's sister flying boat *Beachcomber*. She was now owned by the Science Museum, and was en route from Lee-on-the-Solent to her final resting place, the newly-built Hall of Aviation in Southampton.

It was sad to see her making her final voyage in such an undignified manner. I was saddened, too, at the thought of how Ron Gillies had been financially ruined through his efforts to save the old flying boat. Although the Science Museum had acted honourably and paid a fair price for her, neither Ron, nor several of his friends who had also put cash into the project, ever saw a penny of the purchase money.

Early in 1983, several lads from the local Air Training Corps started helping in their spare time. Three of them, Ross MacFarlane, Graham Nears, and Robin Jones became very involved, spending almost all their weekends and school holidays with us. Today they all earn their livings as aircraft engineers. The boys were put to work going through the interior. Every piece of trim came out, every removeable panel was removed. For most of it, it was obviously the first time out since fitting at Rose Bay in 1964. There was inevitably a lot of corrosion, and some repairs needed. But most of the corrosion was light, and with treatment and re-protection would be good for many years to come. The boys were soon expert at corrosion treatment.

Not only did the trim have to come out, but every cabin window needed replacement. At some time the windows had been liberally coated with paint stripper, causing chemical crazing. Not only did this make them opaque for the passengers, it weakened them structurally. Although a Sunderland is not a pressurised aircraft, the security of its windows is still vital to its safety. If a window was broken by a heavy sea, it would probably lead to the loss of the aircraft. Making and fitting a full set of windows, with the equipment we had available, took very many man hours.

Yet another job was the radios. The light aircraft type radios which had been fitted had proved to be less than adequate. Benny Lynch, the radio man who had come to the rescue in Bermuda, started coming down to Calshot at weekends. His task was to install a complete new radio fit, using second hand units, but of current airliner standard. This was quite a task, as it involved the fabrication and installation of new and complicated wiring looms.

The work described above took us right through 1983, and most of 1984. Then, in October 1984, came a major problem. Our landlords at Calshot, the Hampshire County Council, wanted to turn the site we occupied into a marina. Sunderland Ltd received a letter asking us to move out. It was a blow, but by some remarkably good luck, there seemed to be an alternative available. We had recently been contacted by one Malcolm Moulton, of the Medway branch, Royal Aeronautical Society. Would we be interested in moving the flying boat to the Medway area?

I was despatched to Kent, to investigate the possibilities. The main force behind the move to get the boat to Kent was GEC Avionics. They occupied what had been Short's factory at Rochester Airport. Nearby was the old Seaplane Works, on the River Medway. This was the real birthplace of the Sunderland flying boat, although ML814 had not been built there.

Government cuts had recently brought about the closure of the area's largest employer, the Royal Naval Dockyard at Chatham. The navy were moving out, and the 700 acre site divided. The larger part was handed over to English Estates, a redevelopment agency. But the older, largely Georgian section of the yard, was to be turned into a 'living museum'. GEC Avionics hoped to be able to keep the Sunderland in the old dockyard, and I was taken to see the site.

The main gate was certainly imposing, a brick arch between two massive towers, with a huge George III coat of arms over the arch. Inside, as we drove down a hill to the river, was the dockyard church on our right, and beyond it a large and imposing residence. It was obvious that the navy had not skimped on its architecture in previous centuries. To our left, were three dry docks between the road and the river.

After the dry docks came a row of five covered ship-building slips. These, it was explained to me, originated in the days of wooden ships, when the weather could rot the timbers of a ship even before it was built. The first, and oldest, of the covered slips was entirely constructed of wood, and cathedral-like in its proportions. Next came three intermediate iron slips, and at the far end of the row was the newest and largest, No. 7 Covered Slip. It was here that they hoped we would be able to keep the Sunderland. Open where it fronted onto the road, the great span of this cast iron-framed shed was reminiscent of that of some of the London railway stations, and it dated from a similar period — the early 1850s. Inside, the concrete and stone slip ran downwards to wooden gates at the far end, which held out the waters of the River Medway. Flanking the main slip were two roofed side bays.

Impressive it was, but would the Sunderland fit? Beyond No. 7 Covered Slip was the Open Boat Slip. Between the two were several railway sidings. There was no doubt that the Sunderland could be brought out of the river on this open slip, although it would need a few alterations. The intention was that it would come up the Open Boat Slip, and then be hangared in the landward, level section of the Covered Slip. But on returning to the Covered Slip, and measuring it, I discovered that it was just 80 feet across. The Sunderland was 89 feet long, with 112 feet wingspan. We could not get it in.

Travelling back to Southampton by train, I turned the problem over in my mind. As the two main beaching gear legs of the Sunderland castored, not only could it be moved forwards or sidewards, it could be crabbed at any angle in between. Back at Calshot, I drew and cut out a scale plan of a Sunderland on card. Then a plan view of the Covered Slip to the same scale. I tried tracking the card 'Sunderland' in and out of the 'Slip' at different angles and felt sure that it could be done — but with only inches to spare.

Arrangements were made for the move. We were still far from completing the Certificate of Airworthiness work, but we could easily get the boat into a flyable state and make the move on a Certificate of Fitness for Flight. This would mean crew only aboard. There were hitches. Ross and I worked right through one night to have the boat ready for launching in the morning, only to have a main gear tyre suddenly blow out at 5 am. There was no spare, so another day was lost.

Eventually we had her launched and fuelled. It had been decided that there was no point bringing pilots half way across the world every time we needed to fly, and a further search had been made in England. This led to Captain Ken Emmott, recently retired from British Airways. Captain Emmott had flown Catalinas during the war, and Hythes and Solents with BOAC after the war. But it was now well over thirty years since he had flown a large flying boat.

The CAA was happy to allow Captain Emmott to fly the Sunderland, once he had got in some hours beside a pilot currently on flying boats. To achieve this, Edward Hulton brought over Captain Reg Young from Canada. Captain Young had never flown a Sunderland, but earned his living flying Martin Mars water bombers on forestry protection. Like the Sunderland, the Mars was a World War Two veteran, but in size could dwarf even a Sunderland.

As it was now over two years since the boat had flown, it was desirable to have a test flight prior to departure for Chatham; this would also give the pilots an opportunity to get the feel of the boat. So, on a grey and slightly misty Saturday morning, 17 November, we taxied out into the Solent for the flight. Captains Ken Emmott and Reg Young were up front, with John Land as flight engineer.

The morning's circuits and splashes (flying boat equivalent of circuits and bumps) passed off without a hitch, and it certainly lifted the spirits of all the team who had been working on the boat at Calshot. For most of them it was the first time that they had seen her fly. The crew seemed to have no trouble handling her. I think that there were quite a few surprised spectators around the shores of the Solent that day, and I noticed that on each of the ten or so circuits, we brought the traffic to a standstill on the foreshore road near Gurnard, on the Isle of Wight.

For me it was back to sleeping aboard. Not a very attractive proposition at this time of year, especially as I now had a comfortable caravan ashore. But knowing what I did of the history of disasters to Sunderlands at their moorings, I could not rest easily unless aboard.

A cheerful Captain Ken Emmott, following flight tests from the Solent, November 1984.

(Photo Duncan Cubitt)

If there was a mishap, perhaps I could save her, perhaps not — but at least I stood some chance. If I was sleeping ashore, I probably wouldn't even know of the problem until she was washed up on the beach, or on the bottom.

ML814 was now wearing a new name. Since our arrival at Calshot, Bill Mares had suggested that the boat be renamed Sir Arthur Gouge. Gouge had been the designer at Shorts responsible for the Sunderland, although in fact he had left the company and joined the opposition (Saunders-Roe) before ML814 was built.

I think Edward Hulton was rather reluctant to change the name, as Maureen O'Hara had been keen for *Excalibur VIII* to be retained in memory of her husband. However, as the boat was going to the Medway area, where Gouge had been Chief Designer at the Seaplane Works, I was asked to put the name *Sir Arthur Gouge* on the side.

Just prior to departure for Chatham, a logo appeared on the fin for the first time. A Fleet Street gossip columnist had published something about the flying boat moving to Chatham, referring to it as a white elephant. Someone had said "Why not put one on the fin." Edward Hulton gave his approval, and Ross cut out two big elephants from a sheet of Fablon. We went out to the boat on a wet and rough afternoon, rigged the fin ladder, and the elephants were applied.

On the day scheduled for the flight to Chatham we had to cancel, due to poor visibility. Although the aircraft was equipped for IFR flight, we obviously needed good visibility in the take-off and alighting areas. We finally got away on Tuesday 20 November. Quite a crowd gathered to see us off, and I began to realise how many very good friends I would be leaving behind. From a personal point of view I was not looking forward to the move.

Arrival on the Medway. Flying over the Chatham Dockyard.

(Photo François Prins)

Once airborne, we made one low pass over the old base, then set off eastwards along the coast. Visibility was good, we stayed well below 1,000 feet, and had an excellent view of all the South Coast towns. In fact if at any stage we were not sure where we were, it was usually possible to read the road signs. I remember seeing clearly the damage to the Grand Hotel at Brighton which had recently been bombed during the Conservative Party conference. At Hastings, we turned inland to cross to the Medway Estuary.

After circling the alighting area, which was well down river in Long Reach, we flew up along the Medway at low altitude. The first landmark I recognised was the dockyard, our new home. Immediately beyond the dockyard came the two adjacent towns of Chatham and Rochester, with the river winding through them both. Then on a small hill Rochester Castle, and below it the old Short's Works.

Although I had never seen the Seaplane Works before, it was instantly recognisable. There were the great doors fronting onto the river, which formed the backdrop to so many photos of 1930s and 40s flying boats. The giant letters reading 'Short Brothers Aeronautical Engineers' no longer graced the building, but otherwise it looked unchanged. Through those doors had once rolled some of the most famous passenger aircraft of their time, in the days when Shorts were one of the world's leading aircraft manufacturers.

And from the river below us, all of these great flying boats had first taken to the air. A modern motorway bridge, however, now meant that this section of water was not usable by us. We left the river to make two low passes across the nearby Rochester Airport. It was a 'thank you' to our friends at GEC Avionics, who had been instrumental in getting us here. Accompanying the Sunderland at this stage was a light aircraft belonging to GEC, sent to film our arrival. It was from this small

At our mooring on the River Medway, against a backdrop of the Chatham Dockyard covered slips. No 7 Slip, which we were to occupy, is on the left.

(Photo The Times)

airfield that the first of Britain's heavy wartime bombers, the Short Stirling, first flew and then went into production. A considerable part of the Stirling utilised Sunderland components, in particular the wings and tail.

Then, with most of the inhabitants of the Medway towns well aware of our arrival, we returned to Long Reach. There, we made the usual imperceptible Sunderland touch down (at least with all the pilots I have flown with). If it wasn't for the sudden noise of water against the hull, sounding as if you had landed on a pebble beach instead of the sea, there would be no indication of the moment of contact.

Now came the long, half-hour taxi up-river to the dockyard, one of a number of difficulties in using Chatham as a base. We picked up the mooring perfectly, made fast, and cut the engines. Thankful that this bit had gone smoothly, because we were watched by a welcoming party on the river bank, I opened the forward hatch. Where was the boat to take us ashore or bring them aboard? We looked across at them, they looked back at us. There was no boat.

I had liaised with Medway Ports Authority about laying our mooring, which was just as ordered. For other arrangements, Malcolm Moulton had assured me that he had many old Shorts staff amongst his team, who knew all about handling flying boats, so we would be in safe hands. Thus assured, I had not bothered to go into such obvious details as needing a boat to get ashore.

Eventually, we put our rather dubious inflatable out through a hatch, pumped it up, and ran the crew over to a large boat which was carrying a television crew. It went off up-river with them to the nearest landing stage, some distance away. They were due at some sort of civic reception.

For me it was the usual routine — storm pendant, covers, and so forth. But here there was nowhere ashore to go, so I spent the afternoon taking in the scenery, if it

The Royal Engineers bring the boat into the slip at Chatham.

(Photo Kent Messenger)

could be called that. To one side of the river were the corrugated iron covered slips, to the other side mud, and beyond the mud, a plant for unloading aggregate from dredgers. A less inviting marine environment it was hard to imagine.

Soon after the others had departed it started to rain, and did not let up. Around dark I decided that it was time to eat, launched the kayak through the rear hatch, and paddled up-river to Chatham. The rain did little to improve my view of this dreary town, as I left the kayak by a derelict jetty and walked through a back street (called Ship Lane, if I remember), to the High Street.

Here at least there was a welcome fish and chip shop, but imagine my thoughts when I discovered that I had come without any money. That was enough for me for one night. I paddled back to the Sunderland and climbed into my sleeping bag, cold and hungry.

My first few weeks at Chatham was the only time I have felt thoroughly miserable when working with the Sunderland. It rarely stopped raining, and I was missing having friends around as at Calshot. Perhaps most of all I was missing being in a pleasant location. People tried to help, I was invited out for meals, but was reluctant to leave the boat untended for long.

The elements were no problem here, it was a sheltered mooring with no rough seas, which could be a hazard at Calshot. But it was a narrow river and, of necessity, the mooring was virtually in the shipping channel. There wasn't much shipping, and it was only small coasters, but they were quite big enough to sink a Sunderland. Barrier Reef Airways had lost a Sandringham that way at Hamilton Reach on the Brisbane River.

The tail trolley emerges from the water as the boat is hauled up the slip.

(Photo Chatham News)

By the weekend, gear was starting to arrive from Calshot, and there was work to do. We were given tremendous help by the army, 24 Field Squadron of the Royal Engineers to be more precise. There was a flood wall to demolish, an earth ramp to build, gates to move, a slipway to clear of rubbish — all before we could attempt getting the boat ashore.

When all seemed ready, we fitted the beaching gear. Another problem. Our beaching gear had been badly rusted, and this set had just been rebuilt. A new attachment socket did not correctly align with the pin protruding from the flying boat's hull. Ken Emmott had come down to help, and together we spent several hours one night, leaning over the side of the White Boat and taking turns at filing underwater. Finally, at around 10 pm, and feeling thoroughly frozen, we got the pin located.

Our first attempt at beaching had to be called off. I misjudged things, and the tide threatened to push a wingtip onto one of the brick walls which bordered the lower end of the slip. The second attempt, two tides later and just after dark, was more successful. With a combat support boat on our bow and a tractor on our stern, the Royal Engineers got her safely onto the slip.

Then, when they had her just above the water line, she blew another main gear tyre. I set out on an all-night drive to Calshot to collect a spare, with Malcolm Moulton coming along to share the driving and stop me going to sleep.

It had been Thursday evening when she came out of the river; by Sunday we were ready to put her in No. 7 Covered Slip. This was my moment of truth. All the work of the past weeks was pointless if my assurance that the boat would fit in proved to be incorrect. As the level inside the slip was lower than the road in front, not only did we have to take her in at 45 degrees, but run her down steel ramps at the same time.

With a tractor on the bow and another on the stern, to pull her in, and two more outside to steady her down the ramps, the Royal Engineers eased her, crabwise, into the slip. The manoeuvre took hours, with frequent adjustments to the angle setting of her wheels. As expected, there were only inches to spare as her bow, stern and wingtips passed the support pillars, but she went in. I could breathe again. ML814 was safely hangared in the Chatham Historic Dockyard.

A tight fit, but safely inside the Chatham Historic Dockyard's No. 7 Covered Slip.

(*Photo* Chatham News)

Chapter ten

Dockyard Days

With the move behind us, I could now concentrate on the main goal, getting that Certificate of Airworthiness. Although No. 7 Slip was open to the weather at the front, we were far better off for working on the boat than at Calshot. Since the start of this century, the slip had been used for the construction of submarines, 58 having been built there. Thus there was much useful equipment around which we were able to take over and put to use.

The Medway branch of the Royal Aeronautical Society called for volunteers to assist with work on the Sunderland, and quite a number of keen helpers came forward. Predictably, the enthusiasm soon waned, but we were left with three volunteers who regularly came along to work one day a week. Two others, now retired after a lifetime in aviation, came more often and were to be the mainstay of the workforce throughout our stay at Chatham.

Albert Lock had started his career as an apprentice at the Seaplane Works in the thirties, and ended it as a licensed engineer with Dan-Air. He could put his hand to anything, but was unmatched as a 'tin basher'. When he had started at Shorts, the most complicated compound curvature panels were all hand-worked, an art lost today. His skills were invaluable to us.

The other of the two 'old boys' (they hated this term, but were invariably given it) was Bob Woollett. Bob had started in the RAF as a Halton Apprentice, was a flight engineer on Sunderlands during the war, and worked at Hawkers after the war. Like Albert, he could put his hand to anything on an aircraft, and was an excellent precision engineer.

Albert and Bob's first major job at Chatham was to make and fit a complete new set of flight-deck windows. An aircraft trimmer was employed to replace all the cabin trim which had been ripped out at Calshot. There was still plenty of the never-ending work of corrosion treatment, and our Calshot team of Ross, Robin, and Graham, who made regular visits to Chatham in their school holidays, were masters at this work.

Shortly after our arrival at Chatham, I had the Department of Social Security chasing me. When I had started with Sunderland Limited, it had been explained to me that, as I was an Australian and the company was registered in the Channel Islands, they did not have to make tax deductions. Apparently they had been wrong. The company had to pay the National Insurance contributions owing, and I was hit for three years' back income tax. It wiped out my savings. Working on flying boats might be fun, but it wasn't proving very profitable!

We were still continually rectifying problems caused by previous poor workmanship. One major item was the widespread use that had been made of Cherry Max blind rivets. The type of rivet used had a steel mandrel, perfectly satisfactory in a landplane, but of course on a flying boat the steel rusted and the rivet fell out. We replaced hundreds of them.

By the end of September 1985, we had reason to feel happy — the end was at last in sight. The boat was virtually ready for launching, and in a fit state (we hoped) to

Albert Lock and Bob Woollett pictured on the flight deck.

(Photo Kent Messenger*)*

get its Certificate of Airworthiness. We were still way behind on paperwork; for example the updating of the electrical manual was proving more time-consuming than I had expected because of all the system changes. On top of that, the CAA would not accept the Operations Manual which Antilles Air Boats passed on to us, so that would have to be largely rewritten.

But those two jobs could follow, as with the boat now ready we were keen to get on with the flight testing. There would be one or two preliminary flights to check that we were happy with all the systems, then we would carry out the test flight for the issue of the C of A. The Royal Engineers were called upon once again, as all the performance of getting her into the covered slip was carried out in reverse, and the boat launched.

Ken Emmott was of course now cleared to fly as captain, and on 29 October 1985 he took her up for the first flight since arrival at Chatham. First officer was Gary Wrathall, an ex-RNZAF Sunderland captain who, by a stroke of good luck for us, happened to be working in England. We had only been airborne for fifteen minutes or so when John Land, the flight engineer, beckoned me. "Doesn't look good", he said, indicating one of the oil pressure gauges. "Pressure has dropped a little on number one."

John monitored it closely, and it continued to drop. Then he noticed a hint of an oil temperature rise on the same engine. John discussed it with Ken on the intercom. They decided to shut down and feather the engine. That was the end of the test flight — and it was back to the Medway on three engines.

The long up-river taxi was rather a trial for Ken without No. 1 engine, as the outers are the primary engines for taxiing. Differential use of the engines provides steering, but the inners have little effect. Finally we were back on the buoy, to be told of a further problem. A considerable amount of fabric was flapping loose on one elevator. It had not been a good day.

Taxiing back up the Medway after the disappointing test flight of October 1985. No 1 engine is feathered and fabric is missing from the port elevator.

(Photo Denis Calvert)

The engine problem was bad news, but was just one of those things. The elevator, however, looked like a black mark against those of us responsible for maintenance. We should have detected any weakness on our inspections. Examination showed that, although 99% of the fabric was sound, there was one narrow strip along an underside rib where obviously water had been trapped — rotting the fabric and making it quite weak.

Alwyn Roberts pulled the filter on No. 1 engine, and sure enough there was bad metal contamination. It was most probably a master rod bearing failure, and dropping oil pressure with increasing temperature are the classic symptoms of this. Alwyn had taken over most of the responsibility for engineering since our arrival at Chatham. He currently worked nearby at Biggin Hill, and had previously spent a period working on the flying boats with Antilles Air Boats.

There was nothing for it now but an engine change, so it was quite a set-back to the programme. It would be months before we would be flying again, and immediately ahead of us now were all the problems of putting the boat back into No. 7 Slip. An engine change to us was a major operation, yet to an airline it was only a few hours' work. The difference was that airlines carry what is known as a power plant or QEC unit. This consists of the engine plus all its accessories, mounting frame, and the enclosing cowlings. To change this unit merely requires undoing four large nuts, various hoses and electrical connections. The old power plant is craned off, the new one on. On a Sunderland this can even be done afloat, using a special lifting frame which bolts to the wing.

We were going to have to remove the power plant, strip it, obtain a new engine, build that back up into a power plant, then fit it to the wing. At least the lifting bit would be easy. No. 7 Slip was equipped with travellers (overhead cranes). Although they had not been in use for many years, they were an irresistible challenge to our Calshot lads. They soon got two of the travellers operating again.

Alwyn Roberts changing an engine, using the Sunderland's own lifting gear. The overhead crane which we were later to use can just be seen.

(Photo Graham Playford)

We had an inspector along, he advised us on a few incorrect adjustments and the finer points of operating them, and insurance was organised. The travellers were to prove invaluable for engine changes and many other jobs — I imagine we had the only aircraft repair facility in the country with a full overhead crane system.

The other job resulting from the test flight was of course the replacing of the fabric covering on all the control surfaces. Off came the elevators, ailerons, and rudder; each to be inspected, repaired, and carefully protected against corrosion, prior to replacement of the fabric. It may seem unusual to have fabric on a large and relatively modern aircraft like the Sunderland, but it was normal practice of the period. In the late 1930s, the era of fabric aeroplanes was not long past, and rudders, ailerons, and elevators were normally constructed with a metal leading edge, but with the remainder fabric-covered. The DC3, and most bombers of the period, used a similar design.

On this occasion, due to all the other work we had ahead, the control surfaces were sent away to a specialist historic aircraft restorer, Skysport Engineering in Bedfordshire, for re-covering. On other occasions we re-covered surfaces ourselves.

Work at this period wasn't limited to that resulting from the test flight. Our Dan-Air friends were back with us — there was further plate replacement needed on the hull. After the engine failure, we removed the No. 1 oil tank for cleaning. It proved to be in rather poor shape, so they all came out. I found myself spending a lot of time on the rewriting of the Operations Manual.

Alan Bennett-Turner fitting an outer fuel tank.

(Photo Ian McIntyre)

At this stage Sunderland Ltd was being put under increasing pressure to quit the Historic Dockyard. Although local opinion welcomed us, those controlling the dockyard did not. On arrival we had been offered only a six-month free stay. At the end of that time the chairman of the Chatham Historic Dockyard Trust, Sir Steuart Pringle, announced his intention of "encouraging the Sunderland to fly away" by imposing a rent, and steadily increasing it.

I could sympathise with Sir Steuart's view that a Sunderland had no place in the Historic Dockyard, but in the short term there seemed little point in forcing it out. In those early days of the dockyard's opening to the public, there was little for visitors to see, and the Sunderland was quite popular. We went to considerable effort to put up display material on the history of flying boats in the region. We have been gone from Chatham for over three years now, and No. 7 Slip is still empty. Had Sir Steuart been gentler towards us, the dockyard could still be receiving our rent!

By July 1986, the boat was ready once again for launching. Once afloat, as always, the problems began. To do even a simple job on a flying boat at the mooring, which might only take half an hour ashore, is likely to take all day when afloat.

At Chatham, access was particularly bad. Although our mooring was just off the covered slip, there was no access to the river there at most states of tide. Thus, to get a mere stone's throw to the Sunderland involved travelling half a mile by vehicle to the nearest access (Thunderbolt Pier), then back down-river by boat. It was here that our four youngest volunteers proved invaluable. Members of a local unit of the Royal Marines Cadets, they were soon very skilled at handling the White Boat, and could look after all the fetching and carrying.

A social get together of the team at Chatham.

(Photo Ian McIntyre)

Taxi trials showed up several minor but frequently occurring problems, the exactors and the CSU electric heads. Two exactors and one CSU needed to be changed before we would be able to carry out a test flight. Exactors had always been a headache to Sunderland maintenance engineers and pilots. They were a proprietary make of hydraulic engine control, frequently used on marine craft. When the pilot moved his throttle lever it moved an hydraulic piston, which was connected by a small-bore pipeline to another piston operating the engine's throttle lever. An excellent system when working well, it was, however, prone to leaks which caused the controls to 'creep' in flight.

The constant speed unit (CSU) is a simple flyweight governor which, by controlling oil flow to a piston in the propeller hub, adjusts the blade pitch — automatically giving constant propeller RPM. The pilot controls the governor, thus selecting his desired RPM, by moving a lever. On the Sandringham, this lever was connected by cable to the engine.

In the early days of Sandringham operation in Australia, Stewart Middlemiss got fed-up with his pitch selection altering, due to the cable control and the flexing of the wing. Then he discovered that Liberator bombers, also with Pratt and Whitney engines, had 'electric head' units to control the CSU, the pilot operating an electric switch.

The RAAF had umpteen surplus units, so Middlemiss acquired a pile of them, and they were fitted to the flying boats. This was great while they were new, and if one failed he could afford to throw it away and replace it with another new unit. But we were 35 years on, with very worn-out units, plus several worn-out spares. To make matters worse, they sat on the reduction gear housing, on the very front of the engine, where they were thoroughly sprayed with salt water during every take-off and landing.

We had no written data on the electric heads, and couldn't find a company anywhere in the world willing to overhaul them. Eventually, I made contact with Harvey Lippincott of United Technologies at East Hartford, Connecticut. Himself an active flying boat enthusiast, Harvey is the company's chief archivist, and he managed to unearth the necessary servicing data on our units. Armed with this data, we managed to keep the units more or less operational, although unit changes were frequent. The final solution came in 1990, when an American contact tracked down six 'as new' overhauled units in a US Navy store.

July 18 saw us setting off for the test flight, Ken Emmott in command again, but yet another first officer. This time it was Mike Searle, who remained in this crew position thereafter. Mike runs an aircraft engineering business, and frequently flies other historic aircraft.

The flight was virtually a repeat of the previous one. This time No. 4 engine failed, with just the same indications as we had for No. 1 — master rod bearing failure. Out of the water she came. A replacement engine was ordered from the States. Work went ahead with preparation for fitting the new engine, but we also started to do some deep thinking.

Two engine failures on two consecutive flights, and both from the same cause (master-rod bearing failure), pointed to something being terribly wrong. The primary cause of this type of failure was pilot mishandling, using too low a manifold pressure for a particular RPM setting. This was most likely on descents, with the propeller driving the engine. We did not consider this a likely cause, as all our pilots were well versed with these engines. Besides, they always had a flight engineer watching what they were doing from his panel, and he would soon speak up if the pilot allowed such a condition to develop.

The last major overhaul on the engines had taken place six years before, although they had not flown many hours since then. It was by now far too late to raise this problem with the overhauler. The most likely cause of the problem seemed to lay in the fact that, on a number of occasions, the engines had been left for a period of months without being run. Then, when they were started, it was without prior oil priming. This omission could cause oil starvation at the bearing.

These engine problems were of course bad news for Sunderland Ltd. Not only was there the cost of replacements (around $20,000 to have an engine fully overhauled), but there was the additional cost of delays to the programme.

Programme is perhaps a little too grand a word to have applied to the strategy at this time. It was just a case of getting the Sunderland a Private Category C of A as soon as feasible, and then seeing what was offered. It was hoped to put the boat into some type of commercial operation, although to achieve that, it would be necessary to go the further (and extremely expensive) step of getting a Transport Category C of A.

Many suggestions of possible uses came up, but no definite plans were formulated. Amongst possible operational areas suggested at different times were Egypt, the Mediterranean, Kenya, various Pacific Islands, Iceland, Bermuda, Canada, Norway, Ascension Island, St Helena, South Africa, the Seychelles and Australia.

Quite often a visit would be made to the country concerned and discussions held with people there. So many and varied were the plans for use of the boat, that it was a long running joke amongst those working on the Sunderland to ask "Where are we planning to take it this month?".

Late in 1986, with the engine replaced and all ready for flying, a new problem raised its head. We couldn't get the boat into the water. Although we were tenants of the Historic Dockyard, the Open Boat Slip which we used for access to the river, was controlled by English Estates. They had replaced the flood wall across the top of the slip, and refused us permission to install an access gate, even at our own expense.

The argument centred around them wanting some sort of legal indemnity, holding Sunderland Ltd responsible should any land be flooded. The infuriating part of it was that even if the flood wall was in place, there was no flood protection at all 200 yards further up the river, so the same area would flood regardless. It took months to iron out the problem. In the meantime, we couldn't get to the river, so it was the end of our hopes of flying before winter.

Thus we were forced to make an assessment of the overall situation. Time was now dragging on and on in this attempt to get the C of A, despite the major work having been completed over a year ago. What if we started flying next year, and lost another engine? It seemed quite likely. It was decided that we would replace the remaining engines as well over the coming winter. So the two engines which had suffered master rod bearing failure were despatched for overhaul, one to the USA, one to South Africa. There was no longer any company in the UK, or even mainland Europe, who could carry out this work.

The work ahead of us before the next flight wasn't just limited to the engines. Time was passing since the major work had been done in Puerto Rico, and some of the overhaul work done there was now due again. For instance all the flexible fuel and oil hoses, and there were a great number of them, were due to come out for a pressure check. Quite a few needed replacement. We also removed the flaps for inspection, in particular the bearings on which they run.

Over the time we had been at Chatham, Sunderland Ltd had been negotiating with various Australian groups concerning the possibility of returning the boat to Australia. With this in mind, one such interested party sent over an Australian DCA (but now known as the Department of Aviation) surveyor, Frank Shipway. At one stage Frank had worked at Rose Bay, so he knew the flying boat well.

Frank Shipway arrived at Chatham in February 1987, his purpose being to inspect the aircraft thoroughly, and report on the possibility of re-issue of an Australian C of A. From his past involvement, he knew of trouble areas, of which we had been completely oblivious, and went straight to them. One was some corrosion on a tailplane carrythrough spar, on which we then undertook repairs.

The other main problem he pointed out was the port side lower spar boom, with its history of corrosion going back to the New Zealand days. Following Frank's guidance, we discovered that there had in fact been a repair done on the spar, but that it bore no resemblance to the approved repair scheme for that section of spar. We ended up carrying out a major repair in the area, which hopefully has eliminated the problem for ever.

I think that Frank was pleasantly surprised at the general condition of the boat. After so many years out of airline service, he had expected it to be in a worse state of repair than it was. He was, however, particularly critical of our arrangements for its maintenance.

He had, in fact, caught us at a bad time. Alwyn Roberts, who had been responsible for our maintenance, had recently left to take a position with Shorts in Germany. So I was left trying to run things — with limited experience and no licence. Shortly afterwards, the company took on a new licensed engineer, Don Daniels.

The outcome of Frank Shipway's visit was that he could not see why the boat would not still be able to operate commercially in Australia. Many of the Air Navigation Orders had changed since the boat had left Australia, and these would all have to be complied with before it could fly there again. But provided this was done, there should be no great problem.

Around this time, I had advertised in the local paper for more labour to speed work up. The chap I took on, Alan Bennett-Turner, had not worked in aviation before, but was quite enthusiastic. In fact he was so enthusiastic that he wanted to set up a formal support group to raise funds for assisting with the Sunderland. I had suggested this when we first brought the boat back to England, but seeing that the idea was not appreciated, had let the matter drop.

Alan, however, wasn't so easily put off. He distributed circulars promoting such a group, contacted interested individuals, and so forth. Such assistance was not appreciated by Sunderland Ltd; so much so that after Alan had been away for a period, I was not permitted to re-employ him.

Alan wasn't the only party trying to assist us at this time. Unfortunately, Sir Steuart Pringle was reinforcing his policy of encouraging the Sunderland to leave his Historic Dockyard. To counteract this, there was quite a strong local movement lobbying to keep the Sunderland in the area. Malcolm Moulton, representing the local branch of the Royal Aeronautical Society, discussed the matter with councillors. The outcome of this was that the Medway City Council made an offer to Sunderland Ltd, to provide them with a hangar and slipway for the boat, completely free of charge. We would be able to maintain the boat in the hangar, and fly it as required from there. The only condition attached was that we must guarantee to base it there for at least five years.

The justification for such an offer, using ratepayers' money, was of course the tourist-drawing potential of the flying boat. I couldn't believe our luck. It had always seemed to me that the long term answer for the Sunderland was a museum type hangar in which she could be protected from the weather, maintained, and have paying visitors see her. Each summer she would be able to fly to air shows, and earn income by sponsored appearances, corporate entertainment, and the like. Such a base would ideally be in an area with historic flying boat connections, such as the Medway or Calshot. Hopefully it would grow eventually into a flying boat or marine aircraft museum, as there is currently no such museum in the UK.

The offer was put to Sunderland Ltd but was refused. I just could not believe it and felt very discouraged. Malcolm Moulton was left to rue his wasted efforts and apologise to those he had troubled for help.

Sunderland Ltd must have paid dearly for not accepting this offer. The five years are now well passed but it has continued to cost us heavy rental payments for outside storage plus the additional costs of deterioration and damage, which would have been avoided had we been in the hangar at Chatham.

By September 1987 we were close to being ready to fly. Unfortunately, because we had tackled so much work, it was now much later in the year than I would have liked, but nothing could be done about that. We even gave the boat a new coat of paint to freshen it up, as it had been looking rather patchy. This wasn't a proper repaint, which involves a total stripping to bare metal, but at least it now looked a lot more respectable. If she didn't now complete her test flying programme and get that C of A, she never would. There were four new engines on her, and virtually every system had been renewed, overhauled, or inspected and found good. Ron Gillies arrived from Australia to help with the flying.

Monday 5 October saw us bringing the Sunderland out of No. 7 Covered Slip and parking her on the adjacent hard standing. Only a few minor jobs to complete now, and any necessary engine adjustments to make. You cannot test-run a flying boat's engines on a mooring, so it is essential to have everything completed before launching. We were now working towards a particular date.

October 16 was the 50th anniversary of the first flight of the prototype Sunderland, which had taken place a few miles up-river from us at Rochester. We wanted to make a flypast at the old Seaplane Works on the anniversary. Then, as usual, a temporary problem set our schedule back. An incorrectly connected hose caused some damage to an engine oil tank, and the tank had to come out for repairs. By the 15th it looked like just a few more days' work before we were ready to launch. Ron Gillies was doing engine runs to help our electrician, Ron Albutt, sort out a problem with the cylinder head temperature gauges. Bob Woollett was replacing a tail navigation lamp. Finishing touches were being put to the repaint job — in fact the boat was looking almost like new.

Sadly, we were not going to be flying for the 50th anniversary on the following day. Instead, we decided to have a bit of a party to celebrate the event. Don Daniels' wife, Joyce, and our part-time secretary, June, were going to bring in the eats.

At about 1 am. on the morning of the 16th, I was woken by the sound of wind. It didn't take long to realise that this was no ordinary wind. No particularly strong winds had been forecast, so I dressed to dash out and see that no work stands had been left where they could blow onto the aircraft.

All seemed well outside, except that the inner bracing wires of the starboard float had snapped, allowing it to swing in the wind. I was puzzled as to how this could have happened but, realising that the float attachment points would be strained, I got some ropes for a temporary lashing. I pulled a stand over to the float and climbed up to attach the rope. While I was standing on the float, it suddenly descended the six feet or so until it hit the tarmac, then shot back up — with me still standing on it. The opposite wing was lifting! Now I knew how the bracing wires had broken, and here was I worrying about the minor detail of a loose float.

I dashed back into the covered slip, to find the electricity now cut off. With a torch, I collected some rope and strops from our marine equipment store, then brought them out in the Land Rover. I parked the Land Rover under the port (lifting) float, put a strop over the float and around the front bumper. That should stop the port wing lifting. I went back inside to collect more webbing strops, to do a better job of the fixing.

As I came back out around the corner of the covered slip, I could barely believe my eyes. The Land Rover was laying on its side, the wing was 40 or 50 feet up in the air, and the opposite float was being smashed to bits on the tarmac. The force of the wind was such that the wing had lifted the 1·5 ton Land Rover into the air, allowing the strop to slip off its bumper.

The morning of October 16th 1987. Staff arrive and survey the scene. The Land Rover, which had been lashed under the float to prevent the wing lifting, was rolled on its side despite the weight. The stern had been rolled off the tail trolley and sits on the ground.

(Photo Chatham News*)*

There was little I could do now but wait, hope the wind would ease, and then re-lash the float. For the first time, I started to take in what was going on around me. The wind was howling and roaring. Above its noise, I frequently heard a 'whoosh' in the darkness overhead, like the sound of passing shells I had heard on war films. After some time, I realised the cause of this sound. Large sheets of corrugated iron were being torn from the roofs of the covered slips, flying over the Sunderland like missiles, and disappearing into the distance. The wind was such that the boat's propellers were turning steadily despite the resistance of her 1200 hp engines.

Because the Sunderland was now sitting on only one set of wheels, she could rotate easily, and swung the 20 degrees or so to bring herself head into wind. In so doing she rolled off the tail trolley, smashing much of her rear keel. The starboard float didn't last long, and she dropped onto the wing tip. This soon bent up, and more of the outer wing gradually gave way, working in towards No. 4 engine. I had visions of the boat flipping right over onto her back.

Eventually, after something like twenty minutes, the wind eased, she dropped back onto both wheels, and I fixed her again to the Land Rover. I stayed watching until dawn. The wind gradually lessened, there was no more real damage done, but there was probably little I could have done anyway.

That morning the whole country, particularly the south, was in chaos. Few of our work force were able to get in, but they had guessed the worst. When Ron Gillies arrived from his guest house in Strood he said: "When I heard that wind last night I knew what I would find here today. I have been through quite a few good blows in various parts of the world, and that was equal to any of them."

The comments didn't help my spirits. The boat should have been picketed down and, had she been, she would have survived. I was annoyed at there having been no forecast of excessive winds, the more so when I learnt that they were forecast across the Channel in France. But I still felt guilty. I should have had her picketed.

The bent-up starboard wing is plainly visible in this front-on view.

(Photo Ian McIntyre)

Before long, we had the inevitable press and television callers. How bad was the damage? Could it be repaired? Would it fly again? At that stage, we had not carried out a thorough survey, but I had taken a quick clamber around inside the wings and the bilges, and hadn't found anything too frightening. Yes, I said, I thought it could be repaired and fly again. But whether it actually would, that was another matter. That would depend on available funds, insurance payouts, and so forth.

All of this would take time to sort out. It was obviously the end for the present. Ron Gillies booked his return flight to Australia. Regular work on the boat finished for everyone except myself. It was 50 years to the day from the first flight of the first Sunderland — what a way in which to celebrate it.

We put several large dockyard timber blocks under the wings and lashed her down to them in case of further storms — shutting the stable door after the horse had bolted. We inhibited the engines, just in case we might want them again some day. We jacked the bent and torn stern and put it back on the tail trolley. With little else to do, I spent several weeks repairing the Land Rover, as it had not been covered by insurance.

By early December, it was decided that I might as well remove the badly damaged starboard section of the wing, although no long term decision had yet been made regarding the way forward. Although a Sunderland wing is not designed with detachable sections, as on some aircraft, the four spar booms did join with bolted joints just outboard of the outer engine. Thus, by de-riveting all wing skins in this region, then unbolting the spar joints, it was possible to detach the outer section. Bob Woollett came back in to help me, and after about ten days of outdoor work in a bitter December wind, perched high up in the air, we were able to crane off the mangled piece of wing.

Over the next few months, very little happened. Sunderland Ltd was still weighing up the possibilities available to them. The insurance situation was not good. Not very long before the storm damage, the insurance policy had been changed to one known as a 'component parts policy'. This divides the aircraft into its various components

Work starts on rebuilding the ribs of the damaged inner wing. The outer section is removed.

(Photo François Prins)

(2 wings, 4 engines, 2 floats, etc), and gives each a percentage value of the total sum assured. So if a wing was reckoned to be worth 5% of the total value of an aircraft, and the aircraft was insured for say £1,000, the maximum payout in respect of a damaged wing would be £50. In the case of the Sunderland, damage was limited to one wing, one float, and a small part of the hull. Thus, the payout was limited to a small percentage of the total insured value.

The company took a quote for repairs from Aviation Traders of Stansted Airport, and the quote made the insurance payout look laughable. This was despite the fact that the cost of damage repairs was way below the total insured value of the aircraft. Things did not look bright. The cost of repairs was going to be too great, as I had feared. The boat was put up for sale, as she stood. It was announced in the aviation press that the Sunderland was available for £200,000.

Interested parties started to come forward, including the Medway City Council, who were considering it as a museum piece. Several Australian groups again showed interest, and the possibility of shipping it out there as deck cargo on a container vessel was looked at. From Ireland came an enthusiast, Margaret O'Shaughnessy, who wanted to start a flying boat museum. Margaret had certainly done her homework, was very knowledgeable on the subject of flying boats, and spoke of getting backing to purchase the Sunderland. "We will never see her again," I thought. "Some hope she has of getting backing for a scheme like that."

We pottered on with a few minor jobs, but without any real commitment. The slightly-damaged port float was brought into the workshop in February, and 'the old boys' started repairs on it. The boat was still sitting outside, looking sorry and neglected with a great chunk of wing and both floats missing.

John Hart and Alan Noon rebuilding the damaged stern area. Riveting can involve many hours of work in cramped positions.

(Photo Ian McIntyre)

Then, in May, the whole situation was transformed. Edward Hulton decided to take a gamble. The company could not afford to have the boat repaired by an outside firm. Selling it as it stood would only return a fraction of its real value. Sunderland Limited would tackle the repairs themselves. All the old team came back together, soon to be enlarged. Everyone was smiling. The Sunderland would fly again, none of us had the slightest doubt now. Scaffolding went up under the inner section of the starboard wing, which had also suffered considerable damage, mainly to the skin.

Bob and Albert started the tricky repairs on the many damaged trailing edge ribs. Two of the Calshot 'boys' (Ross and Graham), now well into their aircraft engineering courses, came along for their college vacation. They removed the starboard flap and started repairs on it. Mark Burgess, one of our regular helpers from the Royal Marine Cadets, had just left school, so we gave him a job. I went on a re-equipping spree, as we needed an efficient sheet metal shop to handle the coming repairs. I bought a second-hand power guillotine, and borrowed a large folder. From old dockyard scrap I fabricated a set of joggling rolls and a dimpling press.

Several chaps from Dan-Air started coming along at weekends to fabricate new wing skins, but progress was proving too slow. We took on two additional full time sheet-metal men, John Hart and Alan Noon. Both men, we were to learn, had served their apprenticeships at Blackburns, one of the old British flying boat manufacturers, and a company which had licence-built many Sunderlands.

The Royal Engineers return the boat to No. 7 Slip for fitting of the new outer wing. This photograph clearly shows how the Sunderland had to be crabbed into the slip and down steel ramps, with little clearance on the nose, wing tip and tail.

(Photo Ian McIntyre)

Over the summer months working outside was pleasant, and progress quite good. Making up many new skins is not fast work. Those for the wing were large, 8 feet by 3 feet. Firstly of course, the alclad sheet must be cut to size and temporarily fixed in place. Then it must be drilled out, hundreds of holes to allow for riveting it to underlying frames, intercostals, or ribs, and to adjacent skins. These holes must normally be back drilled, as the structural members are already in place and still have the old holes in them. One wrong hole and it means starting all over again, perhaps days of work wasted.

Preparing to crane an outer section of wing into place.

(Photo Ian McIntyre)

Next the holes need 'dimpling' — press countersinking, so the rivet heads will be flush on the surface. Where skins lap, the inner sheet of the pair must be 'joggled'. This means putting a crease in it, to set it back along the lap, so that the joint will be flush on its outer surface when riveted up. By now the skin will have been on and off the aircraft a number of times. Then, when finally ready to rivet in place, off it comes again for corrosion protection.

Firstly all skins, in fact all sheet-metal components, are sent away for anodising, which is the most effective form of protection available. We were very fortunate in that GEC Avionics at Rochester Airport had a treatment plant, and carried out a large amount of anodising for us free of charge. The large wing skins, however, were beyond their capacity, and had to be taken to a company at Twickenham. When back from anodising, there were then two different primers to be applied to every detail part, before it was ready for final riveting. Mark soon became an expert with the spray gun, and looked after this work.

As the inner wing neared completion, John and Alan started repairs on the damaged rear keel. The hull was supported on a specially made trestle to allow this. One major problem on this repair was the sternpost. This is the large alloy casting which forms the knife edge of the rear step, and incorporates the rear towing eye. A new one was needed. By chance, a small business, Medway Heritage Foundry, had recently been set up in the Historic Dockyard. Its two proprietors had previously worked with a firm producing aviation castings. Subject to certain controls, the CAA agreed to allow the casting to be made by the foundry. This was of great assistance, as to have gone to a large approved company for a one-off job like this could have been prohibitively expensive.

Bob Woollett checking dimensions on new spar tubes and the machined end caps which rivet into them.

(Photo Ian McIntyre)

While all this work proceeded, plans were still being made for the solving of our biggest problem — replacement of the outer wing. The original spar structure was smashed beyond repair. The Sunderland's front and rear wing spar was each a girder, composed of a tee section boom top and bottom, joined by vertical and diagonal tubular members. The girders tapered from four feet deep at the wing root to six inches deep at the tip.

The tubular vertical and compression members, with machined end caps for bolting to the boom, were relatively easy to replace. We manufactured dozens of them ourselves in the course of the job. But the tee section booms were a different matter. Massive at the wing root, they tapered continuously until they looked like a piece of aluminium window frame at the tip. To add to the problem, they were not right angle tees, and the angle varied along the length. Shorts had installed a very special milling machine for their manufacture. No doubt new ones could be manufactured, but enquiries indicated that the cost would be prohibitive.

Even if we could have found manufacturers for all the components, we would then need a jig in which to build up the wing. So we looked into the alternative, obtaining a section of an existing Sunderland wing, which would probably mean doing a deal with a museum. The Southampton Hall of Aviation and the RAF Museum Hendon were not interested. The Musée de l'Air were happy for us to have their outer wing, then we discovered that their Sandringham had been blown against a building, and the starboard wing was as badly damaged as ours.

I had reckoned that I knew the location of every remaining Sunderland, or bits thereof, but as we dug deeper some intriguing stories came to light. We were told of a complete Sunderland sitting in a shed at Entebbe Airport. Further enquiries proved this to be a false lead. I obtained information on the state of the hulk which the RNZAF had abandoned on the Chatham Islands. Although some of the hull still remained, apparently local farmers had utilised the rest for building sheds.

Albert Lock rebuilding the wingtip from Duxford. The patchy skin has been previously repaired with body filler, and must be replaced.

(Photo Ian McIntyre)

Next, the company looked at the possibility of getting the outer wing from the Sunderland at Duxford Airfield, owned by the Imperial War Museum. This was yet another boat returned to England after service with the Aéronavale. Don Daniels and I had paid a visit to Duxford in May to inspect the wing, and its spar appeared to be in very good condition. We were warned that the wing had suffered much damage over the years, and that many of the original skins had been replaced in commercial aluminium.

A meeting was arranged with David Lee, Deputy Keeper at the museum, and Norman Harry. Norman was technical adviser on the restoration of the museum's Sunderland, having started his career as apprentice at the Seaplane Works, and later worked in the drawing office there. An arrangement was made whereby Sunderland Limited would purchase the museum's outer wing; then meet all the costs of refurbishing our damaged wing to a good static exhibit state, and refitting it to the museum's Sunderland.

This gave Sunderland Limited the means of returning ML814 to a flying state, whilst also giving the museum considerable funds towards the completion of their restoration project. In October, Mark and I loaded up our tools, and spent five days at Duxford removing the outer wing. We did it with great care, not only to avoid damage to our 'new wing', but also because we expected to have the task of refitting the repaired wing at Duxford. In the event, this was carried out by a local engineering contractor.

Back at Chatham, I arranged the return of the Sunderland to the covered slip. She came back in on 3 November, courtesy of the Royal Engineers. Our friends at 24 Field Squadron by now felt quite involved with the Sunderland, and were glad to know she would take to the air again. I brought her in the opposite way around from previously, to put the damaged wing on the more sheltered side.

With the boat inside, I had a scaffold platform erected to provide a work area around the wing. Once the 'new wing' had been craned into place and supported on trestles, the platform was roofed with plastic sheeting. We now had a cosy, heated workshop, enclosing the wing, in which work could continue throughout the winter. John and Alan from Hull rejoined us, and soon work was going full swing. They started at 7.30 and worked until around 10 pm, usually seven days a week. I was working almost as long, but unfortunately was not paid by the hour. Occasionally I felt a bit discouraged, when I realised they earned three to four times as much as I did.

Once the replacement section of wing had been paint-stripped, it was only too apparent that much of its surface had been bodged up, as we had been warned. A large proportion of the skin was soft commercial aluminium sheeting. The leading edge and tip must have been badly dented at some stage as the surface was now largely car body filler. All of these areas had to be replaced. As the skins came off they revealed that all the trailing edge ribs were cracked — apparently the wing had been dropped.

At least we had our spar, although even on that a number of the tubular members needed replacement. Availability of materials became a major problem in this work. Realising that it was going to be difficult, I prepared a letter detailing our needs, and sent it to every approved materials supplier and aerospace company in the country.

The Kingston and Brough divisions of British Aerospace managed to supply several different tee extrusions we needed for the trailing edge ribs, for which we were very thankful. I suspect the materials had been in their stores since before the Second World War. However, on the various diameters and gauges of tube we needed for the spar tubes, there was no success.

We got around the problem by importing tubes from the USA. These, however, were not to the same specification as the originals, and every substitution of non-original materials had to be justified. Luckily, we held a Sunderland Type Record, giving the loadings and reserve factors for all structural members. Our aeronautical engineer, Eric Niedermayer, was kept busy with the necessary calculations.

Over a period of months, both Eric and myself were regularly chasing Jack Harbottle or Chris Hookham, of the CAA's Design and Manufacturing Standards Division, with our queries. Their very considerable assistance over this time was a major factor in enabling us to carry out the repair work. The design division was not our only contact with the CAA. Throughout the repair work, we had regular visits from our area surveyor, Tony Holman, to check and advise as the work progressed.

By the end of May, progress was looking very good. There was still some work going on with the leading edge and tip but, apart from that, the wing was almost complete. The aileron had been completely smashed to bits, so we had refurbished a spare. Don and Joyce Daniels were now starting to fabric cover it.

Finishing the wing wouldn't, however, be the end of matters. There were still two floats to finish and rig and, as was to be expected, more hull repairs due to the ongoing effects of corrosion. After all the work, the aircraft would have to be re-weighed. Then, of course, there were the engines to be brought back into operational condition.

For some time now, those of us working on the boat had been hearing rumours about it going to Ireland once it was flying again. In fact back in January we had

Reskinning and replacing some ribs on a section of leading edge.

(Photos Ian McIntyre)

received a visit by two chaps from the Irish airworthiness authority. Then, in late March, came a visit from Brian Taylor, engineering manager of the Irish airline Ryanair. Early June brought a more positive turn of events. Brian Taylor came down again, and went through a list of actions that he wanted me to take to speed up work. He explained that a new flying boat museum was being opened at Foynes in Limerick, the old departure port for the trans-Atlantic route.

Margaret O'Shaughnessy, who had visited us in 1988, now had her museum. Backing had come from none other than Guinness Peat Aviation. Ryanair was intending to operate the Sunderland from Ireland in conjunction with Sunderland Ltd, these operations to be tied-in with the museum. The big day was to be 8 July, fifty years to the day since the first Pan American Clipper had touched down at Foynes. Maureen O'Hara would be there for the opening ceremony of the new museum. And, it was hoped, so would the Sunderland.

What chance of having it over in Ireland by early July, Brian asked? Possible perhaps, I thought, but only with a lot of luck. If much more labour was put on, I thought that we could probably just have her ready to fly by then. If all our work turned out to be 100%, with no problems, then we might make it.

But after so much major work — a rebuilt wing, four changed engines, and so forth, this was really asking too much. The likelihood to me seemed that we would have at least some snags, and the time taken to sort these out would mean missing the deadline. It was obvious from his response that Brian was going to put on the pressure to have her ready for then.

Despite Brian's request for a speeding up of the work I was instructed to proceed much as before. Next came a phone call from Ryanair — how was I getting on with the changes? I told them of my instructions. I forget the reaction, but they certainly were not happy. Then, undoubtedly, came some behind-the-scenes discussions and

Inside the plastic workshop. L to R : Peter Smith, John Smith, Mark Burgess, Bob Woollett, Ross MacFarlane, Albert Lock, Don Daniels.

(*Photo* Kent Messenger)

I was instructed by the company to work under the direct supervision of Bryan Taylor of Ryanair Engineering at Luton Airport.

So it was on! Now there was the prospect of a deal with Ryanair. For the last month or so, I and everyone else working on the boat had been hearing rumour after rumour, trying to piece together the various happenings and visits, but always unsure of the real position or the future.

From a personal point of view, it was probably good news. To be involved in some form of operation under the wing of an established airline was really just what was needed. However, to those who had worked for us for many years at Chatham, some of them as volunteers, the news could hardly be worse. After all those years of dedicated involvement, now that the boat really looked like flying properly for the first time, it was to fly away.

Further phone calls followed from Brian Taylor. He spelt out his plans in detail. I was to spare no expense in speeding up work. Before long, he would send down some of his staff to de-inhibit and check the engines. But before that, he would engage an experienced trouble-shooter to come down and "work with me", to get things rolling.

Several days later, Brian's 'man' arrived. He was a tall, wiry, ex-sergeant major type. He strode in through the workshop, which was the access to my office, totally ignoring everyone in it. In the office he asked: "Are you Mr Smith?" "Yes." "I've come to help you. I am what the Americans call a ramrod. Now where is the paperwork?" The name certainly suited. From then on he was Ramrod.

What followed over the next four weeks, I would much prefer to forget. Ramrod's first action was to contact the local employment office. Soon we had about eight new workers on site. I was required to use these men to carry out aircraft maintenance

Joyce Daniels at work on the fabric covering of an aileron.

(Photo Peter Smith)

work. Quality control was going to be much more difficult and I was soon tearing my hair out. I wasn't the only one upset. The first to go were John and Alan. Certainly, it wasn't far from the end for them at any rate as their contracts were now almost completed.

They had been keen to see it through, and see her fly. They had been the mainstay of our labour throughout the storm repair work, and had proved to be hard and extremely skilful workers. John and Alan were worth more to me than the eight new men who had been brought in, but I made no attempt to talk them into staying. I could understand how they felt.

Next, Mark wanted to go. I had two long talks with him to get him to stay. At present, he was indispensable to me. He knew the ropes, knew how jobs should be done, and knew our stores system. He was continually in demand by the newcomers, was run off his feet, and often found himself supervising a team of older men. To make it worse, they were all on about twice his rate of pay. I knew the feeling!

Ramrod spent a large amount of his time on the phone and was ruthlessly efficient at getting what we wanted, when we wanted it. We were continually needing various components or materials to complete the work, and in the normal course of events many of them would have taken days if not weeks to arrive. Ramrod continually harassed suppliers on the phone, until in desperation they would deal with his demands to get him off their backs. No expense was spared, and items were often collected by motor cycle messenger. This way of doing things was a whole new world to me.

Although we had this extra labour, there was a shortage of access towers and other equipment. Then, one of the new workers was killed in a fall, while washing the tail. It cast a very dark shadow on the whole operation. Although he had not been our employee, Sunderland Limited was held partly responsible in court, as the accident occurred on our site.

At this stage, with work nearing completion, there were panels and covers galore to be refitted. With unskilled people on it, there was infinite scope for damage. The Sunderland contained a confusing mixture of hardware. Originally all threads had been British, but after being modified and worked on around the world there were now probably more American than British fittings on her. We had comprehensive stocks of AGS parts, but to expect someone who had never before been close to an aircraft before to know the difference between a 2BA and a 1032 screw, was just asking too much.

After about a week, I could take no more. I had been working long hours and seven days a week for months now, and this was the final straw. I contacted Ryanair to say I was leaving. The reaction was quick, Brian Taylor rode down on his big Honda. "I know you are finding things difficult at present," he said. "But just think, it won't be for long. In a few weeks you will be in Ireland, and be your own boss again. Once there, we will need you, and you will move onto standard Ryanair rates of pay."

He was of course right. What was a few more weeks of this. And the thought of that pay — it would mean at least a 50% increase in my basic rate and overtime.

Soon the Ryanair engineers came down from Luton to work on the engines. I didn't have to worry about them, they knew exactly what they were doing. They de-inhibited the engines, the boat was wheeled outside, and engine running started.

Two painters from Ryanair also paid a visit to Chatham. On went the company's logo, and yet another change of name. This time the boat became *Spirit of Foynes*.

Just before launching the boat, we carried out a most unusual exercise. A compass swing was due. This is a problem with a Sunderland ashore, as it has large steel legs bolted to the side, which won't be there when it is flying. We normally carried out an air swing; the RAF had used special wooden cradles in which they sat the hull, then removed the legs. To Ramrod these were minor hindrances. He had the Sunderland towed around a piece of dockyard waste ground by a steam traction engine, carrying out the swing as he went. It looked like something out of Jules Verne!

Now we were ready to launch and Ramrod handed this operation over to me. We were due to go in on Monday 3 July, but strong cross-winds on the slip forced me to cancel. This was bad news, as the coming Saturday we were due in Ireland. On Tuesday, the winds were even stronger. I decided to go for Tuesday night's tide (actually 0154 Wednesday morning).

Everything was prepared before dark, and some floodlights hired. How I cursed those tall walls either side of the slip. As the evening wore on, the wind dropped away as I had hoped. By midnight there was barely a breath. As a precaution, I had two lines from the tail, one each to a handling party on either bank of the slip. Down she went at high water, and a launch took her out to the mooring — without any problem.

The first job now was fuelling. With the Sunderland afloat, Stephen and Mathew — our two youngest cadet volunteers — had the chance for a little revenge. On his arrival, Ramrod had banned them from going near the Sunderland, although for years they had regularly assisted with all the unpleasant jobs — washing the boat, tidying the workshops, sweeping the covered slip.

Now that there was possibly some reward for all their work, seeing it flying, they were told that they weren't wanted. But with the Sunderland on the water, they were again in demand. Ramrod found himself dependent on their services as boat handlers, just to get himself out to the Sunderland.

The problems of flying boat maintenance. Changing an elevator on the Chatham mooring.

(Photo George Prager)

The first of many. Ken Emmott is invited to make a pass down the runway at Gatwick Airport.

(Photo Alan Timbrell / Gatwick Airport Ltd)

By Thursday, all was ready for a test flight. It really would be a test after so much major structural work since the last flight. Crew were Ken Emmott and Mike Searle, with Geoff Masterton as flight engineer. This first one would just be a shakedown flight, seeing all systems were functioning correctly, checking handling. Later would come the full test flight to a laid down schedule, leading to issue of the C of A.

The day we should have departed for Ireland. Edward Hulton sits on a flood defence wall, while work nears completion on changing a cylinder on No. 3 engine.

(Photo Kent Messenger)

After taxiing down to Long Reach, Ken made several fast runs on the step. All seemed well with systems and handling, so he lifted her off. She was airborne again. In the months of indecision following the hurricane, I had doubted we would ever see this day. It was just six days under three years since ML814 had last flown.

First, Ken did a few local circuits, then headed for the Kent coast. We had only been airborne for ten to fifteen minutes when a bad oil leak developed from under No. 3 engine. It wasn't sufficient to make us curtail the flight, provided it didn't worsen. It would, however, need sorting before we could leave for Ireland or obtain the C of A. During the flight, Ken was invited to make a low pass along the runway at the RAF and civil airfield at Manston. It became the first of many. Over the coming months, and years, the greatest hazard for Ken when flying the Sunderland, became the barrage of requests for 'low passes' over airfields. Amongst those visited have been Gatwick, Heathrow, London City, Hurn, Farnborough; not to mention a host of RAF stations.

For those of us involved in the storm damage rebuild, the flight was the proof of all our work. We were over the moon — she had flown perfectly. For Ramrod, the flight was a major disappointment. He had undertaken to have the boat in Ireland in two days time (the 8th). There was an oil leak to solve and the C of A test flight yet to conduct.

We knew where the leaking oil was coming from, the rear or accessory section of the engine. Although this particular engine, No. 3, had now been fitted to the Sunderland for several years, this had been its first flight. There had been no signs of the problem during ground runs.

A high speed test run on the Medway Estuary.

(Photo Kent Messenger)

A heavy radial piston engine consists of a number of sections, one behind the other. At the front is the reduction gear and propeller. Next is the power section, the ring of cylinders around the crankshaft. In the case of our R1830s, 2 rings each of 7 cylinders. Next back is the 'blower', a big fan driven from the crankshaft to supercharge the engine. At the rear is the accessory section, a casing full of gears from which are driven the magnetos, pumps, generator, and so forth. The electric starter contains its own reduction gear, and is bolted to the very back end of the crankshaft, where it can engage directly with end of the crankshaft via a dog.

As the leak was from the vent pipe of the accessory section, it seemed that either this section was pressurising and blowing oil out, or the scavenge system which returned the oil to the oil tank was not functioning correctly. Don Daniels telephoned Fields in South Africa, the company who had overhauled the engine. They suggested that the problem was likely to be caused by a bad fitting of the starter dog.

But this idea was discounted and a compression check was made. One cylinder appeared to have no compression. By now it was Friday evening, so a team was organised to carry out a cylinder change overnight. I was despatched to pack and get some sleep. After all, I would be moving home tomorrow. I didn't bother to pack as I knew we were going nowhere. I was convinced that a cylinder change would not have the slightest bearing on our problem.

Next morning the job was completed, but unfortunately the oil leak was still there. There was absolutely nothing wrong with the old cylinder, the compression check must have been taken with the valves open. By now we were inundated with press and television reporters, awaiting the widely advertised departure for Ireland. A fast taxi-run was carried out on the Medway to placate them. We now sought other causes for the leak. The oil pump was changed. We rigged up a modified vent system which someone had told us would solve the problem. It wasn't until the following

A happy day. Edward Hulton holds the newly issued Certificate of Airworthiness, with Don Daniels looking on.

(Photo Ian McIntyre)

Sunday that the starter was finally removed. There was the problem, obvious once we had the starter dog off. A bronze sealing ring, between the dog and the end of the crankshaft, had not been properly located during the overhaul. Exactly as Fields had suspected, and advised us over a week ago. This meant that oil from the oilway in the crankshaft was steadily flowing out the rear end. However, it took quite some time to fill the rear casing and then flow from its vent, so had not shown up on ground runs, or when the overhauler had the engine on his test bed.

We proceeded with a more organised test flight programme, as the panic was now over. Don Daniels was busy preparing the masses of paperwork, covering years of work, for final perusal by the CAA. On 2 August the Certificate of Airworthiness was issued. It was now ten years since the company had first approached the CAA on the possibility of obtaining a C of A for ML814. I would think she holds the record for the longest running saga ever concerning the issue of an airworthiness application.

It was a happy moment for Edward Hulton, and a proud moment for those of us who had worked on ML814 throughout her stay at Chatham. We felt especially thankful to the many people in the Civil Aviation Authority who had put up with us over the years, who had encouraged and guided us when things were difficult, and had made this day possible. We knew that there were many others who, particularly after the storm damage, had believed we would never reach this point.

Geoff Masterton, the flight engineer, at his position in the astro-dome during engine start-up, at Calshot, September 1991.

(Photo Etienne Bol)

Left: Nelson takes in the view on a wet day. He often spent periods of months on end living aboard the boat.

(Photo Peter Smith)

Around the WORLD

This superb study by François Prins shows G-BJHS moored on the Thames during the visit to London, in September 1982, in aid of the South Atlantic Fund.

G-BJHS, carrying the name Sir Arthur Gouge, moored on the Medway in October 1985. Gouge had been the designer at Shorts responsible for the Sunderland, although he had left the company to join Saunders-Roe by the time ML814 was built.

Chapter eleven

Ireland and Beyond

At 12.55, on Thursday 3 August, Ken Emmott started the two outer engines, then gave me the all-clear signal to release the short slip line. It ran out through the ring of the mooring buoy, and *Spirit of Foynes* was under way on the Medway for the last time.

Half an hour later, when airborne, Ken made a farewell pass by the covered slip, and we saw a forlorn little group standing outside waving to us. Some of that group had given so much of their lives to help get us flying, and now reluctantly we were leaving them. When Bob Woollett had been asked for his thoughts earlier, he had replied "I feel like I'm saying goodbye to an old friend for the last time".

What a relief it was to be in the air, and for the first time in months able to feel relaxed. Our working environment over those last eight weeks could only be described as a madhouse. And problems, now of a non-technical nature, persisted right to the end. Only two days before the departure, the crew and myself were very doubtful if we would ever leave for Ireland.

We knew that there was some behind-the-scenes bargaining still going on between Sunderland Ltd and Ryanair. Tempers became frayed at our end when the company demanded to know why I had filled all the fuel tanks just for a flight to Ireland. More phone calls to Ireland followed, and the matter was finally resolved when Sunderland Ltd received a fax from Dublin confirming that all the fuel would be paid for by Ryanair

Amongst the passengers on the flight were Mark Burgess and Graham Nears, both (along with myself) to be employed by Ryanair on maintenance. Graham had just completed his final exams, and was now a licensed engineer. Also on board was my pigeon Nelson (what else could you call a one-eyed pigeon, living in a dockyard where the *Victory* had been built?).

Nelson was by now quite a seasoned flying boat passenger, like those of his breed who had been carried on wartime Sunderlands. And I was a very relaxed passenger with Ken Emmott at the controls. There may have been a 40-year gap between Ken's two periods of flying boat command, but it certainly didn't show. His wealth of experience was evident in the way he handled the boat, and I would happily fly anywhere with him.

We followed the South Coast as far as Lyme Bay, passing over the old Calshot base, of course. Then we flew cross country to the Bristol Channel, and west to Pembrokeshire. On reaching Milford Haven, Ken made two passes along the waterfront at Pembroke Dock, a reminder of her wartime days based here with 201 and 422 squadrons. The two big hangars still stood, but were inaccessible now to flying boats as the slipway had gone.

From PD it was Strumble Head, and then across St George's Channel to Ireland. Shortly after 5pm the smooth green countryside below gave way to partly-wooded hills — we were nearing Lough Derg, our destination. Ken had visited the area previously to familiarise himself with the terrain, and brought the Sunderland low over the Arra Mountains to alight on the southern end of the lough. He taxied across the broad expanse of lough until it narrowed down to again become the River

Shannon. On the river, just above the village of Killaloe, was Derg Marine, our new base. As we neared the marina came the first sign of a welcoming committee — the first signs even that anyone had observed our arrival.

Four or five little inflatables, each manned by a small boy clutching a noisy marine hooter, sped in alongside the Sunderland from where their welcoming hoots were quite audible, even above our engines. This unusual flotilla escorted the flying boat down to the marina, where we picked up our mooring across the river from the rows of pontoons and pleasure craft. There was a minor hitch ten minutes later, when the Sunderland floated gently back across the river until it was in the midst of the pontoons and small craft — our newly-prepared mooring apparently happy to float along with us.

Putting the aircraft on short slip to depart for Ireland.

(*Photo* The Times)

The biggest inconvenience of this mishap was that we had to remain on our own anchor for a day or so, until a new mooring could be laid. This meant the unwelcome burden of anchor watches, although with three of us sleeping aboard it wasn't too onerous. Conditions seemed so gentle that I was not unduly concerned, although weather can always change.

Our arrival at Lough Derg was deliberately low key; no press, no television, no official party. We were to remain 'in hiding' here until the following Monday, when we would fly down to Foynes for our official arrival in Ireland. With no work needed on the Sunderland, a few days to relax and take in the surroundings were most welcome. We were guests of Brian Cullen, who ran the marina. His main trade was in the hiring out of holiday cruisers, and he operated a tiny single-engined amphibious flying boat to provide rapid transport, should one of his cruisers have a breakdown.

Brian was also a leading figure in the establishment of the Foynes Flying Boat Museum, hence our being based here. Charlie Blair had been the first to bring a large flying boat onto the lough (in 1976), and Ron Gillies spent the winter here with *Southern Cross* in 1980/81. And of course *Excalibur VIII* (as she then was) had refuelled here the same year, en route to Southern France. Lough Derg offered a wide expanse of fresh water, and the shelter of rolling hills to all sides. A more suitable location for the basing of an elderly flying boat would be hard to find. And the more I saw of this beautiful corner of Ireland and met its inhabitants, the more I knew that I would be very happy to make Killaloe my home.

Anyone to go ashore? There was no shortage of ship to shore transport in Ireland.

(Photo Peter Smith)

Two days after our arrival, I borrowed Brian Cullen's van and made the drive down to Foynes, mainly to check out the mooring arrangements and chat with the crew of our safety boat there. The launch crew took me out to inspect our mooring, which was prepared just as I had requested. They normally handled the mooring of large vessels arriving at Foynes, and were obviously skilled boatmen. It was good to know that should anything go amiss on our arrival, we had a reliable crew standing by.

The visit was also a chance to meet Margaret O'Shaughnessy again, and see her museum. Located at the waterfront end of the village's rather dispersed main street, the museum occupies what looks like just another stone house. But this unimposing building was once a convenient terminal for the transiting passengers, the jetty being directly across the road. The building, with an open central yard, lent itself well for conversion to a museum. There was a small cinema, a tea room, displays of charts, radios, and other artefacts. The only thing missing was a flying boat, but we were about to deliver one.

On Monday afternoon we set out for Foynes. I was in the bow compartment, and Ken was on the point of pressing the starter button for No. 1, when a tiny inflatable came skating under the wing and stopped alongside. Ken leant out of the flight deck window, about to make some appropriate comment. The nine-year-old stalled his engine, shouted "Have a good trip, mister!", pulled his starter cord, and was off.

For the short flight down to Foynes we had a number of Ryanair people on board, including their chief pilot, Captain Bluey Gardiner, and his family. Bluey had made several visits to us at Chatham, and had been turning on the hospitality now that we were in Ireland. As Ken taxied the boat up to the Foynes mooring and I surveyed the welcome on the shoreline, I was astounded. I had expected a curious crowd, but there was an unbroken line the length of the sea wall, in places three or four deep.

On the mooring at Killaloe, with Derg Marine in the background.
(Photo Peter Smith)

Everyone for miles around must have turned out. As well as our launch, a local tug was standing by, dressed overall.

The VIPs were brought out, including the Minister for Tourism and Transport. After a look over the Sunderland, they and the crew went ashore for further ceremonials, while I stayed aboard to show through the endless crowds of visitors, who were ferried over for a quick look around. Finally our crew and passengers came back out, plus an Irish television crew who joined us for some filming. Ken carried out several circuits and splashes for their benefit, then cut engines out in the channel so they could transfer back to the launch. Once they were clear, he took off yet again and we returned to Lough Derg.

Although this visit to Foynes had been quite an event, and was covered by Irish Television, it was really little more than a dress rehearsal for the coming weekend. Margaret O'Shaughnessy had arranged a much bigger publicity event, with many foreign TV and press people coming along. It was going to be a major event for the new museum. That evening, the flight crew had been invited to a party at P J McGoldrick's house, further up the lough. A short time before they left, a message was brought out to the boat, asking me to go and see Ken in the crew's hotel on the edge of the river. He had just come off the phone, and was obviously taken aback. He had been ordered to leave for England first thing in the morning. Sunderland Ltd and Ryanair had failed to reach an agreement on joint operation.

Ken had explained the local situation, how further events for the museum were planned, and reminded the company of Ryanair's commitment to pay for all fuel, regardless of whether the deal was concluded or not. But the directors were adamant — we had to return immediately. So off the crew went to their party, aware now that it would be their last night in Ireland.

In the morning I went over to the hotel to see the crew, and was invited to join Bluey Gardiner for breakfast. I didn't know how to react to his first question, as it involved giving away the fact that we were leaving. Then he intimated that he already knew, so I could relax. By the time we were ready to depart at midday, all of Killaloe knew what was happening, and there were many goodbyes to say. We had only been here five days, but had made many friends in a that time. The people I really felt sorry for were the individuals like Brian Cullen (and family) and Margaret O'Shaughnessy, who had made so much effort for nought.

There was little talk aboard during the two-and-a-half hour flight. Ken had been told to return to Calshot, although after landing in the Solent we learnt from our boatman, Mr Etheridge, that there was no suitable mooring available at Calshot. As we taxied in, I was saddened to see that a great Solent landmark, the *Calshot Spit* Light Vessel, had been replaced by a mere automated light buoy.

With no mooring at Calshot, we had to make the long haul up Southampton Water to Hythe, where there was a port authority buoy available within the Southampton Docks area. This was Mike Searle's first visit to the Solent by flying boat, and as Ken taxied through familiar waters I could hear him on the intercom, recalling places and events from the past. He told Mike of the time he was first officer on a local training flight in a BOAC Solent. The captain approached too fast, bounced, and the entire forward lower section of the boat was cut clean off. The crew were lucky to escape with their lives. Tales of that nature make you realise that when the wartime generation of flying boat pilots finally disappear, there will never again be that great depth of experience available.

Having moored at Hythe, we had a bit of a wait for customs to arrive, but once they cleared us everyone disappeared rapidly. Graham Nears lived just down the road, Mark Burgess went back to Chatham as he was owed some holidays. It was back to old quiet times, with just myself and Nelson living aboard. When Graham reached home, he rang Ryanair and asked "What about that job you promised?" He was told to report to their Luton hangar with his toolbox, and had his first full-time job. In retrospect, I think I should have tried the same, as so little was to happen with the Sunderland over the following years.

Although I thought we would quickly find a new base, I was soon proved wrong. The obvious course seemed a return to Chatham, where a packing firm was just starting to pack and load our stores and equipment. Enquiries showed that we could go back there, but because of the Historic Dockyard's policy towards us, the company was understandably reluctant to return there.

ML814 was now due for another name change and I quickly stuck on what I considered to be the only name she should ever have carried — *Islander*. She is still carrying it today, although I suspect that it won't be long before it sees another change.

While at Hythe we made several flights for filming purposes. One was a flight around the Isle of Wight for a local TV current affairs programme, in which Nelson starred (I suspect that he is the most photographed pigeon in the world). Then we spent three days filming for an award-winning documentary, "The Last African Flying Boat". Although mainly about a Catalina travelling down the Nile, the film commenced with shots on the Sunderland, purporting to be an Empire Boat. I even located an original British Power Boat Company tender, as used by Imperial Airways, to participate in the filming.

The mooring at Hythe was just off the Southampton Docks. Here the QE2 is departing on a cruise.

(Photo Denis Calvert)

The ironic part of the film was that it opened with the narrator lamenting the fact that there are no longer any large Shorts' flying boats, so he must make his voyage down the old Imperial Airways route by Catalina. All of which is said against a backdrop of modern footage filmed aboard our Sunderland.

The Sunderland's first visit to an airshow was made at the end of August, when we went to West Malling in Kent. Firstly Ken had to practice his 'performance', and qualify for a special pilot's air show endorsement, before being allowed to participate. Ken worked out his air show routine, with assistance and advice from our co-pilot, Mike Searle, himself an experienced air show pilot. The practice flight was carried out over Calshot, with the beach treated as the airshow crowd line. The sunbathers there on one weekday in late August were treated to an unexpected display. By the time he had finished, the perspiration was rolling off Ken, and he had aching arms and legs. Fifteen minutes of 'showing off' an aircraft the size of a Sunderland, with no power controls, is very heavy work.

From all reports, the Sunderland 'stole the show' at West Malling, but little did the spectators realise how lucky they were that we arrived at all. As Ken was carrying out his pre-flight checks on the mooring at Hythe, an exactor suddenly went unserviceable, giving him no throttle control on one engine. Everyone 'had a go' with it, but it just would not bleed up. We had no spare unit available at the time to enable a quick swap. It looked as though we would miss the show. Ken continued to fiddle with the lever, and suddenly it came good. The problem had most likely been a bleed valve not seating properly.

During the weeks we spent moored at Hythe, I had the benefit of a most unusual mail delivery service. Our mooring was just off the end of the long Hythe Pier, in continual use by the commuter ferry service across to Southampton. The mail would arrive on the quaint little train that brought ferry passengers down the pier, its driver giving me a shout so that I could paddle across in the kayak to collect it.

ML814 makes its first appearance at an air show — West Malling, Kent, August 1989. The Ryanair markings have been hastily painted out.

(Photo Denis Calvert)

Once the company had rejected Chatham, I thought we would automatically return to our previous base at Calshot. We started to look at other possible bases in the area — hovercraft slips on the River Itchen, a creek on the Isle of Wight, even Hythe Marina. After checking them all out — and most of them had already been considered and ruled out the last time we were in the Solent area — it was finally decided to beach the boat again at Calshot. She came up the slipway on 28 September.

At least when the hull came out of the water this time, it was still in good condition. Partly this was due to a relatively short period afloat, not much over two months. Also, I had now come to the conclusion that the only suitable finish for a flying boat getting the sort of usage the Sunderland did (long periods afloat with very little flying), was a standard marine finish, including anti-fouling. It had, however, taken quite a bit of trial and error work to find one which would reliably stay on an aluminium hull at 90 knots.

When all our equipment and other gear started to arrive from Chatham we had to purchase a 40 foot container to store it in. But how could everything be stored and still be accessible? The container was stacked full from end to end, with no hope of access. Sunderland Ltd hired a Portakabin, which acted as stores and office. That helped a little, but the container was still too full to allow reasonable access to items stored in it. From having an excellent working base at Chatham, we now had to endure a very poor one here. Each year, when we needed to employ expensive labour to prepare the boat for flying, that labour was much less efficient than it would have been at Chatham and progress can be affected by the weather.

It took the remainder of the year just to get established at Calshot, if established it can be called. All that we managed to do on the Sunderland before the year was out was the necessary washing and protecting immediately she came ashore, and later to get her engines inhibited.

What a year 1989 had been. The first half had been hard, working long hours, but extremely challenging. The repairs to the storm damage had been a big job for us to take on, but we achieved our aim. Then followed the brief, chaotic, and sad period of preparation for departure to Ireland. It went very close to reducing me to a nervous wreck.

After the delight of obtaining the Certificate of Airworthiness, and the high hopes with which we had set off for Ireland, there was now a huge question mark over the future. Almost immediately after our return from Ireland, the boat was put up for sale, and was now in the hands of an aircraft broker. What a disappointing end to what could have been a very worthwhile arrangement. The only redeeming feature of it all, from a personal point of view, was that my home had returned to Calshot. From my desk (where this book was written) I could again look out under the Sunderland wing and over Southampton Water, to the buildings on Hamble Point which once housed the Fairey Aircraft Company. In the opposite direction and across the Solent can just be seen the large Westland hangar on the waterfront at Cowes, previously the flying boat manufacturer Saunders-Roe. Everywhere there are reminders of the great maritime aviation past of the area.

October 1989 — the Sunderland ashore again at Calshot. She is brought up on the northern slipway (just beyond the main jetty, with a hovercraft on it).

(*Photo* Southern Evening Echo)

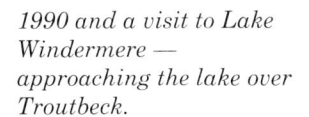

1990 and a visit to Lake Windermere — approaching the lake over Troutbeck.

(Photo Paul Bramham / Westmorland Gazette*)*

Serious work to prepare the boat for flying in 1990 did not start until March, the earliest that continual outdoor work was really sensible. There was the usual round of corrosion repair work to carry out, as well as scheduled maintenance checks. Quite enough to keep four of us busy for several months, and we were joined for a short time by two contract painters, brought in to remove the Ryanair markings.

One total change from our working methods at Chatham was that I decided to make no further use of volunteer labour. In the latter days at Chatham, the directors of the company had complained that the use of volunteers had led to an atmosphere from which they felt excluded. Now that I realised their real attitude to the volunteers, I was glad to be free of the additional burden they created. I had introduced volunteers to try and reduce labour costs. The retired professionals and the youngsters were excellent. The 'weekend volunteers' however, were a mixed blessing. Often they thought that because they were helping, I should look after them. They would expect meals, some even hinted that I should provide them with accommodation. Yet I personally gained nothing from their help — it was the aircraft which benefited. Volunteers had also meant that I needed to work weekends and holidays, for which I received nothing. Now that I was without them, I wondered why I had saddled myself with such additional work in the first place.

By early June the Sunderland was ready to launch, and she went in on the 11th. After refuelling and a test flight, we visited three airshows on Saturday 16 June: Biggin Hill, RAF Coningsby (home of the RAF's Battle of Britain Flight), and the Naval Air Yard at Fleetlands (near Portsmouth). A number of minor snags had shown up on this flight, and we had a busy few days clearing them and refuelling again in preparation for a trip to Lake Windermere. There had been considerable interest over a number of years in getting the Sunderland to the Lake District, mainly because of historical links with flying boats in the area.

Under wartime pressures to increase flying boat production — particularly in areas away from the bombing risk — Shorts had constructed a large factory on the

Touchdown on the lake.

(Photo Paul Bramham / Westmorland Gazette*)*

shores of Lake Windermere. Although it was demolished immediately after the war, many locals remembered the connection, and were keen for a visit by the Sunderland. In fact Albert Lock, who had worked for us at Chatham, had been sent to Lake Windermere from the Seaplane Works, in order to help train the local work force. The visit had been arranged to coincide with the 1990 Lake Windermere Festival, and we made the flight up on 28 June. En route there were, of course, requests from various airports for a flypast; on this occasion these included Boscombe Down and Liverpool.

Ken had made his usual reconnoitring visit to the area beforehand, and brought the boat in between the surrounding hills, to approach the lake through the valley at Troutbeck. Looking down, we could see great crowds gathered on vantage points to watch our arrival, and once Ken had alighted and taxied up to our mooring off Bowness, the Sunderland was surrounded by small craft.

The festival organisers were paying to bring the Sunderland to the Lake District, and hoped to recoup their costs by organising tours to it. A traditional lake launch brought visitors out from Bowness, and while each group was aboard I gave them a guided tour of the flying boat and a brief talk on its history. At £4 a head, they were taking over £1,000 on a good day, and easily covered their costs. By the end of our stay I had lost my voice, but the big surprise to me had been the potential of this form of use for the Sunderland. Several cocktail evenings had also been arranged aboard for local dignitaries, and from the guests' reactions it was obvious that events of this nature could be a very marketable commodity.

A week or two after getting back to Calshot, there was a minor incident which demonstrates how the unexpected is always likely with a flying boat. It was just on dusk, and I was sitting at my evening meal. Glancing outside, I was surprised to see a large bulk carrier, well out of the shipping channel, heading for the shore just ahead of the Sunderland. Dashing to the radio, I was just in time to hear the calm voice of the vessel's pilot, "...letting go my anchors. The bow is now running aground." He certainly was aground, and had only narrowly missed a yacht on the next mooring to our own.

Taxiing to our mooring on Lake Windermere, just off Bowness.

(Photo Paul Bramham / Westmorland Gazette)

Before long the harbourmaster was in touch, requesting me to be prepared to move the Sunderland should it be needed — to allow tugs to refloat the grounded vessel. I telephoned Graham Nears, who drove down to Calshot so that I would have a second pair of hands if necessary. We removed the storm pendant, and stood by ready to move.

The problem, we learnt, had been caused by the vessel losing power to its steering gear, and having no choice but to go aground. She had been leaving Southampton for Italy, with a cargo of English grain. I wondered what might have been the result had the failure come a minute earlier. In the event we did not have to move, and the grainer came off by going astern under her own power at a little before high water, around midnight. Her bow passed close to the Sunderland, swinging on the tidal stream.

Shortly after our return from the Lake District, the C of A expired — it was already 12 months since its issue. The timing was unfortunate, right in the middle of the summer, when the company most expected to use the Sunderland. But with the annual inspections and maintenance work due before the C of A could be renewed, it was back up the slip for the boat. By the time the work was completed, the company was undecided as to whether it was worth re-launching the boat and carrying out the C of A test flight, or whether to lay it up for the winter. Finally, as no other flights were arranged for the year, we decided to leave it ashore and renew the C of A at the beginning of the following year's flying.

An important event in late 1990, at least from my point of view, was that I finally talked the company into purchasing a new (second-hand) set of beaching gear tyres. At last I would be able to move the boat around when ashore without blowouts. It was hardly surprising that the old tyres had given continual problems, they were down to the canvas and badly perished. The only one carrying a date read 'Dunlop—1943'.

In December, we had a visit from Captain Erik Dokken of the Royal Norwegian Air Force. It would soon be the 50th anniversary of 330 Squadron, and they were keen for 'their' old Sunderland to make a return visit to Sola for the celebrations. Costs of the trip were worked out, and Erik returned to Norway to see if he could organise sponsorship to make the visit possible. It would not be a cheap trip, as straight operating costs of the Sunderland, not allowing for overheads, are between £2500-£3000 an hour.

Over the winter of 1990-91, the company laid me off for three months. It was the first time in the ten years that I had been working on the boat that maintenance was completely closed down. Then, in late March 1991, I was asked if I could have the boat ready to fly to Norway in two weeks time. Used as I was to unpredictable demands this request left me incredulous. I knew only too well that following the 'no maintenance' period over the winter, there would be at least some corrosion repair work to carry out before we could launch; and then there was the C of A to obtain. We would be very lucky to do it in two months, much less two weeks. Sadly, the timetable was just too tight and the Norwegians would not get their visit.

Don Daniels at the flight engineer's panel. Note the considerable changes from the panel in its earlier days, as shown in Chapter 2.

(Photo Southern Evening Echo*)*

When work did start there were soon five of us on it, and it was not until early September that we were ready to launch. It turned out (as usual) that there was far more work to do than I had reckoned. There were several repairs to carry out on the planing bottom, a skin to replace under the tail, another on the hull side near the galley hatch, and some structural repairs around some of the cabin windows. A large number of flexible fuel hoses were due for replacement. Three of the propellers had to be removed, for a 'Notice No. 75 check'. This involves stripping them, to check the blade roots for corrosion. At Chatham, the CAA had approved us to carry out this work ourselves. But now, without a propeller shop of our own, the work had to be put out to an overhaul firm.

We finally launched on 9 September, which was only just in time. The company was still trying to sell the boat and had arranged for it to be auctioned through Sotheby's on September 24th. We must have it flying and with a C of A by then.

Once it was on the buoy, there was the usual tedious job of refuelling to carry out. There is a strict limit on the weight permissible when on beaching gear, so not much

Looking forward in the upstairs cabin.

(Photo Southern Evening Echo)

fuel can be put on until she is afloat. We normally used the White Boat for refuelling, carrying two 200 litre drums at a time, and using an electric pump to push it up into the tanks. In rough weather it can be quite a long and difficult operation. Wash from passing traffic can be something of a hazard when refuelling, or carrying out other work on the boat. Although there is a continual flow of large container vessels, tankers, and passenger ships into Southampton, the biggest problem is created by the wash from smaller, faster, vessels.

If this is a real problem, it usually only requires a call to VTS (the port control radio), to request a reduction (slowing down) from passing ships. The nautical flavour given to messages, such as the following, often makes me smile: "Canberra. VTS Southampton. Can you give a reduction passing Calshot. The Sunderland flying boat is bunkering."

The test flight went off successfully on 12 September, and the new C of A was issued on the same day. On the 18th we made a one-and-a-half hour local flight, primarily for a Dutch film crew who were making a documentary. They certainly were not taking any chances of not having sufficient film footage. In addition to their own chase plane, they had a camera crew on the Sunderland (including a remotely operated camera on our wing tip), and another camera crew in a fast boat.

Then on the 24th came the big day — the boat was up for auction. The auctioneers had brought several prospective purchasers down to see the boat prior to the sale, which was held in their London saleroom. Considerable publicity was given to the event in the media, where it was said, without any great foundation, that the expected price was around £1m. I watched the sale on the evening television news, and learnt that the boat did not sell, having failed to reach its reserve price.

The auction was a great disappointment from the company's point of view, with bidding remaining well below the Sunderland's true value when considered against

Steven Evans working on spark plugs.

(Photo Peter Smith)

comparable historic aircraft. Spitfires, which are relatively commonplace, have changed hands at over half-a-million, and think of all the extra 'value for money' in something the size of a Sunderland — quite apart from its uniqueness.

I think that the failure of the boat to sell had a deep effect on all of us at Sunderland Ltd. The boat was left on the mooring until 14 November, and we only made one other short local flight in late October. The company line was for taking it abroad, and they seriously considered moving it to Norway, where there may have been a hangar available. There was still the possibility of free hangarage at Chatham.

From my standpoint, 1991 was a totally wasted year. After all the years of effort we had put in to get that C of A, when we had it, no use was made of it. For the entire year we had made only three short local flights. Following the success of the Windermere visit in 1990, I had hoped for more revenue-earning visits to other suitable locations, but there were none. We were at our mooring only a few miles from Southampton at the time of their big annual on-water boat show. It seemed to me an excellent opportunity to get a large number of visitors aboard and some worthwhile revenue.

Eventually, the decision was made to bring the boat ashore again at Calshot. By the time we got her up the slip, thoroughly washed out and engines inhibited, it was almost the end of the year. Then, just before Christmas, I was visited at Calshot by a fellow Australian.

Various attempts had been made over the years to return *Islander* to Australia, and although the interest had come from a number of organisations, the most persistent was the Sunderland Flying Boat Foundation. This body, formed in 1986, was comprised mainly of former members of No. 10 Squadron RAAF, and existed

Launching at Calshot, September 1991. The boat is lowered down the slip by tractor, then taxied to its mooring.

(Photo Simon Clay)

solely for the purpose of trying to return a Sunderland to Australia. It was anticipated that the flying boat would eventually go on display at the Australian War Memorial, Canberra.

After a number of years spent trying to raise the purchase price of ML814, but without success, the Australians were becoming concerned. It had not sold at the 1991 auction, but easily could have, and would have thus slipped from their grasp for ever. Following the country's close wartime involvement with Sunderlands, and the civil role they played postwar, the foundation members were most anxious to seize their last possible opportunity to have one preserved in Australia.

So in late 1991 the Sunderland Flying Boat Foundation tried a different approach. Bryan Monkton, a member of the foundation, came to England and put an offer to the company. The foundation would lease the aircraft, fly it to Australia, set up an operational base for it at Rathmines — the former RAAF base — and finally hand it back to us with an Australian Transport Category Certificate of Airworthiness.

The scheme had appeal to the directors of the company, and they showed considerable interest in it. They had always hoped for a commercial operation for the boat but, following Charles Blair's experiences, knew that it was most unlikely under the American authorities, and likely to be both difficult and expensive under the United Kingdom's CAA. The Australian authorities, however, were still indicating a willingness to have *Islander* operating again on their register. Sunderland Ltd had, in fact, contemplated taking the boat there on several occasions.

Back in Australia, the foundation set about turning their proposal into a reality. They would obtain the Australian C of A before leaving the UK, and pay for the flight by carrying passengers. Careful costings and route planning were carried out, then advertising for passengers began.

Passing an Isle of Wight ferry while taxiing to our take-off location in Southampton water.

(Photo Peter Smith)

Here is an extract from their brochure: "The last great flying boat voyage from England to Australia by four-engine Sandringham Flying Boat. Departing Southampton for Sydney. This historic flight will be made as an extended (21 day) re-enactment of the early Empire Air Mail Flying Boat Services operated originally by Imperial Airways and Qantas Empire Airways. Ports which will be visited on route are: Marseilles, Augusta, Cairo, Bahrein, Karachi, Calcutta, Rangoon, Singapore, Sourabaya, Darwin, Cairns, Brisbane and Sydney."

The plan was to carry 15 passengers at a fare of £10,000 each, which was reckoned to cover the delivery flight costs. Tickets were sold and deposits taken. Then a successful deal was concluded with the Murdoch-owned News Corp to buy advertising rights for the flight at A$100,000. So apparently the money was there, and at last it seemed a return flight to Australia would be a reality.

The foundation dispatched one of its members, solicitor Bruce Miles, to England by courtesy of Qantas, to conclude the deal. He came to Southampton where I met him and brought him to Calshot for a day — we discussed the plans and he had a look over *Islander*. Then he returned to London for further discussions with Sunderland Ltd.

Around the WORLD

Above: Preparing to put the boat on short slip, River Medway, July 1989.

Below: Islander over the Solent, September 1991.

The Last FLYING BOAT

Above: On the mooring at Killaloe, in Ryanair livery and carrying the name Spirit of Foynes.

Below: Taxiing on Southampton Water.

In the meantime a new player had come on the scene, apparently offering a more attractive deal. This new interested party was the promotional company Air Miles, who were about to expand their operations into the United States. By chance the Air Miles logo depicted a flying boat, thus making the Sunderland an ideal vehicle for a publicity programme.

An Air Miles team came down and spent a day with us at Calshot, looking at the boat and discussing their plans. Their intentions were to charter the boat, fly it to the United States, and carry out a promotional tour of a number of the major cities. They realised that a flying boat's ability to alight, or at least moor, right in the downtown centre of so many cities made it ideally suited for publicity purposes. Air Miles even indicated a possible interest in purchasing the Sunderland should the charter prove successful.

However, Air Miles' need to operate to a tight and totally reliable schedule with a 'one off' aeroplane meant that a lot of spares back-up and operational support was necessary. This caused a rapid escalation of cost estimates, which ended at over $1m (for one promotional tour).

A second major problem for Air Miles was the time factor. They needed to have the boat ready to coincide with their launch date in the US. It was rapidly becoming apparent that their date could not be met. So, with considerable disappointment, Air Miles decided not to go ahead with the promotion.

Later, Sunderland Ltd attempted to resume discussions with the Sunderland Flying Boat Foundation but by then it was too late. After a long battle by the foundation to put their plans in place and get the necessary corporate backing, and then having to call it all off, there was no way that they would be able to re-interest their backers now.

If I had found 1991 a disappointing year, I was finding 1992 even more so. For the first half of the year absolutely no work was carried out on the boat, and she was deteriorating rapidly. During June I was re-employed by the company to carry out some overdue corrosion repair work, although this may have been influenced by my mentioning an offer I had of employment elsewhere.

Meanwhile, although the Australian deal and the Air Miles charter were both off, there were other occasional flurries of negotiation concerning possible uses for the boat. Other large concerns expressed interest in some form of commercial nostalgia-type operation for the boat and there were visits to Calshot to arrange.

Early in 1990, a Kent-based company, Orient Airways (Aviation Heritage) Limited, announced that it was their intention to purchase *Islander* and operate her commercially in the Eastern Mediterranean.

Orient Airways' plan was to take passengers from the Orient Express at Venice, and fly them on to Istanbul via stops at Brindisi, the Ionian Isles, Athens, and the Greek Islands. From Istanbul there was the option of returning to London by Concorde, or returning to Venice by flying boat (via Dubrovnik).

With overnight stops en route at luxury hotels, the plan would certainly have provided passengers with a flying boat experience in the grand style. BOAC had pulled out of flying boats largely because of the cost of maintaining bases right down their route, just for the use of their own fleet of flying boats and yet here was a new company planning on setting up facilities in at least seven ports of call, for one flying boat.

For a commercial operation a minimum of base facilities was likely to be a safety craft and fire tender at each port, not to mention passenger handling and fuelling facilities. Even before operations started, the company faced an enormous problem in obtaining a British transport category C of A for *Islander*.

Despite these problems, Orient Airways started to plan operations in a serious manner. Some long-retired flying boat men were called in as consultants, and a route proving flight was planned. I was even sent a draft employment contract for the quaintly titled position of 'Senior Aero Mechanic'. Amongst the detail of the contract I was amused to find that "...the post, by necessity, specifically excludes your communication with the travelling public unless authorised in writing by the Engineering Director in advance." Understandable, of course, as one could hardly have a crude colonial relating flying boat stories to the passengers!

I firmly believe that a viable nostalgia-type operation could be set up with *Islander*, but only with a careful, down to earth approach and a minimum number of ports of call. As time passed, and proving flight dates were cancelled, it became apparent that Orient had been unable to come to a suitable arrangement with Sunderland Ltd. The auction in 1991 would have been the time to make their move and yet I understand they did not even place a bid.

Not to be deterred, Orient Airways continued to make proposals throughout 1992, and in a press statement spoke of acquiring "a range of Flying Boats, some original vintage types, such as the Solent, Sunderland and PBY 5A Catalina, and some modern types currently in production." Good luck to them, how wonderful it would be to see such an operation, but I am afraid that I remain sceptical concerning its viability.

As 1992 wore on, it became obvious that Sunderland Ltd were anxious to find a way out of the constant financial burden of maintaining the aircraft. The company tried to get a reduction in the heavy rental paid to Hampshire County Council, our landlords at Calshot. It was to no avail however — they looked on the flying boat only as a source of funds, not as an important piece of the region's aviation heritage and a very great potential asset.

This prompted the directors to turn to a council which they knew would welcome the boat — the Rochester upon Medway City Council. They were most certainly still interested in having the boat back in Medway, and a group of councillors and officials made the long journey to Calshot by coach for another look at her.

For a while it looked as if we would be returning to Chatham — I was even asked to de-inhibit the engines and get them running in readiness for a move. It was not to come, however, through financial considerations. The intention had been to return the Sunderland to No. 7 Covered Slip on a twelve month lease\option to purchase arrangement between Sunderland Ltd and the Council. After being recommended by the Council's Policy and Resources Committee, acceptance of the scheme was narrowly defeated at a Council meeting on October 13. Although keen to see the Sunderland back in Medway, councillors felt unable to justify the cost of the lease arrangement (around £70,000) at a time of severe financial stringency. [1]

Shortly after this news, rumours reached me that the boat had been sold to an American collector. I discounted such reports at first, until several days later when Sunderland Ltd confirmed that they had finally sold the boat. The new owner was Kermit Weeks, who had made a name for himself as an outstanding racing and aerobatic pilot, then gone on to establish the Weeks Air Museum in Miami.

Preparing for take-off on a filming flight. There is a cameraman in the bow and a remote controlled camera mounted on the wing tip. Note the flight engineer taking in the flag.

(Photo Etienne Bol)

Kermit Weeks had been showing an interest in acquiring the Sunderland for his museum (which included a number of other flying boat types) for some time, and made a visit to Calshot to inspect it. He had even purchased a property in northern Florida which had a lake large enough for the Sunderland to operate from.

Then, in August 1992, came the news that his museum at Tamiami Airport had been badly hit by Hurricane Andrew, and that a large part of his collection of some 40 historic aircraft had been severely damaged or destroyed. I had assumed that this catastrophe would absorb all his energies for some time, and was quite taken aback at news of the sale.

Through an intermediary in the UK, plans were put in place for the transfer flight of the boat to the USA during the summer of 1993. As I write this there is a considerable amount of work to be undertaken before that can be done, including the renewal of the C of A and the refurbishing and fitting of the long range, trailing edge fuel tanks.

I imagine that the move to Florida will bring about the end of my period of involvement with ML814, a fact I regard with mixed feelings. Undoubtedly I will miss her, but it is perhaps time that I took a new job. Quite possibly I will miss my home here on the end of Calshot Spit, in the shadow of Henry VIII's castle, more than I will miss the Sunderland. For someone with a love of ships and the sea, not to mention aviation history, it is hard to picture a more perfect spot to call home.

So, it would seem, Edward Hulton's period of ownership has also finally come to an end. Had he not purchased ML814 back in 1979, the last flight by a Sunderland would have taken place many years ago — to the disappointment of those who enjoy the magnificent spectacle of a large flying boat in action. Only the company's accountants will ever know how much has been spent on ML814 over those 13 years. It should have kept her flying well into the next century — even had she not been earning an income. One can only reflect on what might have been.

Although the Sunderland has been based in the UK now for over 10 years, what benefits have accrued from its presence? Not many people have had the undoubted pleasure of being been able to see her flying, and only a handful have enjoyed a flight in her. Over the same period the other two large, privately-owned historic aircraft in the country (a Flying Fortress and a Catalina amphibian) have flown regularly and been appreciated by millions — and without massive financial input by their owners. As land-based planes of course they have been much easier to maintain and operate.

Yet because of its British origins, its wartime role in the RAF and the Battle of the Atlantic, and the fact that it is a flying boat, ML814 could have generated far more interest and affection around the country than either of these 'foreigners'. If only people had been given the opportunity to support it, assist with it, or even just see it.

So what will the change in ownership mean? After several years of uncertainty it would now seem certain that the boat will remain flying, and will receive the maintenance that it needs. It is Kermit Weeks' intention to refit the gun turrets and return ML814 to its military layout. This may make it a more interesting museum piece, although I shall regret the demise of *Islander*, both as a unique flying boat and for the reminders of Lord Howe Island days which she provides.

Take-off from Southampton Water, September 1991.

(Photo Etienne Bol)

Never to be seen in this country again? ML 814 — at its first appearance at an air show — passes over a 'foreigner', the B-17 bomber, at West Malling in Kent, August 1989.

(Photo Kent Messenger)

On the other hand, she is bound for a country where, with the exception of her new owners, she may not be fully appreciated. The name 'Sunderland', in a flying boat context, means little to the average American, whereas in Britain it still conjures up heroic memories of the nation's struggle for survival against the U-boat, as well as the links with the romantic era of civil flying boats.

For me personally, however, the saddest note is that *Islander* will not be making her final resting place in the country where I feel she most belongs, and where undoubtedly she would be most appreciated — Australia.

Footnotes Chapter 11.

[1] *Chatham News* — Friday 16 October 1992

Appendix one

Flights made by ML814 with 201 Squadron

Note: This table includes all operational sorties flown by ML814 with 201 Squadron, but not necessarily all other flights, although it is probably close to being complete.

Date	Captain	Details	Times\Duration
(1944)			
24 April	A Poole	Transit: Wig Bay—Pembroke Dock.	Up 1505 Dn 1640
25 April	A Poole	Transit: Pembroke Dock—Calshot.	Up 1240 Dn 1405
9 May	D Easton	Mount Batten: Air Test.	40 mins
10 May	D Easton	Base familiarisation flight.	Up 1035 Dn 1505
10 May	D Easton	Transit: Mount Batten—Pembroke Dock.	1 hr 15
12 May	D Easton	Radar exercise, flare dropping Carmarthen Bay.	Up 2205 Dn 0030
14 May	D Easton	Anti-submarine sweep Bay of Biscay.	Up 1306 Dn 0227
19 May	D Easton	Anti-submarine sweep Bay of Biscay.	Up 0215 Dn 1527
22 May	D Easton	Oasthouse exercise. Submarine attack with low level bombsight.	Up 1320 Dn 1800
29 May	D Easton	Local flying.	2 hrs
30 May	D Easton	Radar homing, flare dropping.	3 hrs 55
8 June	D Easton	Anti-submarine patrol Bay of Biscay.	Up 2055 Dn 1018
11 June	D Easton	Anti-submarine patrol Cork 1.	Up 0559 Dn 1840
14 June	D Easton	Air Test.	30 mins
15 June	D Easton	Anti-submarine patrol Cork 42.	Up 1720 Dn 0530
20 June	D Easton	Anti-submarine patrol Cork 42.	Up 0510 Dn 1807
23 June	D Easton	Anti-submarine patrol No. 83 (Brest).	Up 2023 Dn 0923
30 June	D Easton	Anti-submarine patrol No 105 Recalled: Bad weather at base.	Up 1236 Dn 1731
1 July	D Easton	Air test.	30 mins
6 July	D Easton	Anti-submarine patrol Cork E. Sighted black object (whale?). Bomb carrier gear problem.	Up 1155 Dn 0100
8 July	D Easton	Anti-submarine patrol No 132 Brest patrol. Landed Mount Batten.	Up 2108 Dn 0627
9 July	D Easton	Transit: Mount Batten—Pembroke Dock.	1 hr 45
20 July	R McCready	Anti-submarine patrol. A CLA search SW of Ireland.	Up 1207 Dn 0247
23 July	R McCready	Anti-submarine patrol (HH sortie). Diverted Scillies.	Up 1656 Dn 0707
24 July	R McCready	Transit: Scillies—Pembroke Dock.	Up 1809 Dn 1917
1 August	R McCready	Anti sub patrol Channel approaches.	Up 1555 Dn 0348

Appendix one (cont.)

Date	Captain	Details	Times\Duration
7 Aug	D Easton	Anti-submarine patrol off Brest Peninsula.	Up 0706 Dn 2045
9 Aug	D Easton	Air test	40 mins
10 Aug	D Easton	Anti-submarine patrol Rover 34 off St Nazaire.	Up 2117 Dn 0956
14 Aug	D Easton	Anti-submarine patrol Bay of Biscay.	Up 1935 Dn 0810
18 Aug	D Easton	Air test, Low level bombing.	1 hr 40
20 Aug	D Easton	Anti-submarine patrol Rover 28 Ile d'Yeu—Gironde Estuary.	Up 1706 Dn 0645
23 Aug	D Easton	Air test, bombing, radar check.	2 hrs 10
26 Aug	D Easton	Anti-submarine patrol off St Nazaire.	Up 0617 Dn 1950
31 Aug	D Easton	Locating and escorting destroyers 47°N 09°30W. Rtn Calshot.	Up 1345 Dn 0016
1 Sept	D Easton	Transit: Calshot—Pembroke Dock.	
5 Sept	D Easton	Parallel track search No 6 Bay of Biscay.	Up 0814 Dn 1932
8 Sept	D Easton	Anti-submarine patrol No 29 South of Fastnet.	Up 0836 Dn 1655
18 Sept	A Wilson (Australian)	Anti-submarine patrol South of Ireland.	Up 2057 Dn 0945
24 Sept	L Wilson (Canadian)	Anti-submarine patrol. NW of Ireland—Clyde approaches.	Up 2150 Dn 0850
30 Sept	L Wilson (Canadian)	Anti-submarine patrol NW of Ireland.	Up 2000 Dn 0747
3 Oct	A Wilson (Australian)	Anti-submarine patrol. 50°N 18°30W	Up 1710 Dn 0720

Appendix two

Flights made by ML814 with 422 Squadron

Note: This table includes all operational flights made by ML814 with 422 Squadron, but not necessarily all other flights, although it is probably close to being complete.

Date	Captain	Details	Times\Duration
(1944)			
4 Dec	C Gauss	Transit: Calshot—Pembroke Dock.	1 hr 35
8 Dec	L Giles	Bathmat exercise.	3 hrs 35
9 Dec	A Mills	Base—Angle Bay.	10 mins
10 Dec	L Giles	Low level bombing exercise. Radar homing.	55 mins
11 Dec	A Mills	Bathmat exercise.	5 hrs 15
13 Dec	—	Squadron returns to operational flying.	
28 Dec	L Giles	Anti-submarine patrol English Channel.	Up 0515 Dn 1645
30 Dec	G Maier	Anti-submarine patrol English Channel.	Up 0149 Dn 1656

Appendix two (cont.)

Date	Captain	Details	Times\Duration
(1945)			
6 Jan	L Giles	Air Test.	1 hr 15
7 Jan	L Giles	Low level bombing exercise.	1 hr 50
18 Jan	—	Pembroke Dock hit by bad storm.	
28 Jan	L Detwiller	Transit: Pembroke Dock—Mount Batten.	1 hr 20
28 Jan	C Gauss	Radar homing and bombing exercise (night).	6 hrs 05
1 Feb	L Giles	High Tea exercise.	1 hr 50
2 Feb	L Giles	High Tea exercise.	3 hrs
5 Feb	G Maier	High Tea exercise	1 hr 20
7 Feb	K MacKenzie	High Tea exercise.	2 hrs 30
13 Feb	D Park	Circuits and splashes. Solo test.	1 hr 15
20 Feb	C Gauss	Transit: Pembroke Dock—Belfast.	2 hrs

Appendix three

Flights made by ML814 with 330 Squadron

Note: This table probably shows all operational flights made by ML814 with 330 Squadron from Sullom Voe. There are likely to be a number of flights made after the German surrender which are not recorded as, due to the move from Sullom Voe back to Norway, the Squadron records are not complete for this period. There is no attempt to record the continual transport duties carried out by ML814 from its return to Norway, until it left the Squadron in January 1946.

Date	Captain	Details	Times\Duration
(1945)			
2 May	O Evensen	Anti-submarine sweep NW Shetlands. Screen for Russian convoy JW66.	Up 1300 Dn 0005
4 May	Ö Torgersen	Anti-submarine sweep NW Shetlands.	Up 1320 Dn 0105
7 May	T Bugge	Anti-submarine sweep.	Up 0859 Dn 1700
7 May	—	Sunderland G of 330 takes Allied party to Oslo.	
8 May	—	Formal German surrender.	
11 May	O Evensen	Flew escort to HMS *Devonshire* and convoy, taking Crown Prince Olav and Gen Thorne to Oslo	11 hrs.
13 May	—	Last operational flight by 330 Squadron.	
27 May	Ö Lorentzen	Air Test.	1 hr 10
13 June	Ö Lorentzen	Transport flight: Sullom Voe—Sola.	Up 1132 Dn 1512

Glossary and Abbreviations

AAB	Antilles Air Boats.
ADF	Automatic Direction Finder.
Aéronavale	French naval air service.
Aéropostale	Early French airline started as mail service.
Aérospatiale	French state-controlled aerospace company.
AFBS	Ansett Flying Boat Services.
AGS	Aircraft General Supplies (a British term for aircraft hardware components).
Alclad	Dural sheeting coated with pure aluminium to increase its corrosion resistance.
Aldis lamp	Powerful, hand-held electric lamp, incorporating a telescopic sight.
ANA	Australian National Airways.
Anodising	Protective electrolytic coating applied to aluminium.
ASI	Air Speed Indicator.
APU	Auxiliary Power Unit.
Asdic	A method of detecting submarines underwater by reflection of sound 'pings'.
ASV	Aircraft to Surface Vessel (name for an early radar).
BOAC	British Overseas Airways Corporation.
BP	Burns Philp (A Pacific shipping and trading company).
BRA	Barrier Reef Airways.
Braby dock	U-shaped floating pontoon into which flying boats are winched tail first.
CAA	Civil Aviation Authority (UK).
Capt	Captain.
Chook	Australian term for domestic hen.
C of A	Certificate of Airworthiness.
CLA search	Creeping Line Ahead.
DCA	Department of Civil Aviation (Australia).
Dickie	RAF slang for pilot.
DME	Distance Measuring Equipment.
DNL	Det Norske Luftfartselkap. Norwegian airline later incorporated into SAS.
Dural	Copper alloy of aluminium, far stronger than commercial grades of aluminium.
Exactors	Remote hydraulic controllers used for throttles and other engine controls.
FAA	Federal Aviation Administration (USA).
FPM	Feet per minute.
Fg Off	Flying Officer.
Flt Lt	Flight Lieutenant.
Flt Sgt	Flight Sergeant.
FO	First Officer.
Gee	Wartime radio navigational aid.
Grp Capt	Group Captain.
HF	High Frequency (normally referring to a long range radio).
HF/DF	High Frequency Direction Finding.
HNMS	His Norwegian Majesty's Ship.
IFR	Instrument Flight Rules.
Intercostals	Longitudinal strengthening members running between the frames of a Sunderland hull.
Jump seat	Additional or observer's seat on an aircraft flight deck.
Kts	Knots.
LSA	Lowest Safe Altitude.
Met	Meteorological.
Metox	Equipment fitted to U-boats to detect aircraft radar emissions.
MF	Medium Frequency.
MO	Medical Officer.
Mouse (to)	To lock a shackle pin (or hook) with wire.
MU	Maintenance Unit.
NDT	Non-destructive testing.
NSW	New South Wales.
PD	Pembroke Dock.
Pitot head	A probe protruding forward from an aircraft to measure air speed.
PLA	Port of London Authority.
Plt Off	Pilot Officer.
Porpoising	Violent up and down surging motion which can develop during a flying boat take-off run.
QDM	Magnetic bearing to a ground station.
QM	Quartermaster.
QEC	Quick engine change unit.
RAF	Royal Air Force.
RAAF	Royal Australian Air Force.
RAI	Réseau Aérien Interinsulaire. Pacific subsidiary of TAI.
RCAF	Royal Canadian Air Force.
RFC	Royal Flying Corps.
RNZAF	Royal New Zealand Air Force.
SE	Special equipment. RN and RAF term for radar.
Sgt	Sergeant.
Short slip	Line passed through a mooring-buoy and held by several turns on the flying boat bollard.
Sqn Ldr	Squadron Leader.
Storm pendant	Heavy cable attached just below water, from the aircraft's bow to the mooring-buoy.
TAA	Trans-Australia Airlines.
TAI	Transport Aériens Intercontinenteaux (later UTA). French international airline.
TEAL	Tasman Empire Airways Limited.
TOA	Trans-Oceanic Airways.
VHF	Very High Frequency (normally referring to relatively short range radio equipment).
VTS	Vessel Traffic Services. The call sign used by many port control (marine) radio stations.
Wg Cdr	Wing Commander.
Wimpy	Common RAF term for Wellington twin-engined bomber.
Wingco	RAF slang for Wing Commander.
Wop/AG	Wireless operator / air gunner (also W/AG).

A

Aéronavale 182
Aéropostale 143
Aérospatiale 141, 143, 146, 153
Aggie Grey's 81
Air Miles 209
Air New Zealand 84
Aircrew Association 150
Airlines of New South Wales 97, 109, 110, 112, 115, 122
Airtech Services Inc. 131
Albutt, R W 155, 174
Alcock, George 126, 129
Alcock, Sir John 129
Alena - motor launch 109
Algiers 40
Allais, Capt 87
Alness 71, 73
Amalgamated Wireless of Australia 100
American Export Airlines 120, 123, 142
Angle Bay 17, 23, 24, 49, 53, 54, 57
Anglesey 17
Ansair 95
Ansett Airlines 95, 155
Ansett Airlines of Papua New Guinea 122
Ansett Flying Boat Services 90, 91, 96, 103, 119, 122
Ansett, Reg 90
Ansett, Sir Reginald 113, 119, 120
Ansett Transport Industries 90, 111
Antigua 127
Antilles Air Boats 121, 122, 123, 126, 127, 129, 166, 167
Apia 81
Apollo, HMS 66
Arado 196A floatplane 71
Argentina 65
Ariadne, HMS 66
Arra Mountains 193
Asdic 12, 57
Astro compass 137
ASV 46, 54, 69
ASV radar 14, 16, 22, 75
Athenia, SS 13
Athens 209
Auckland 84
Auction of Islander 205
Augusta 208
Australian Army 113, 115
Australian Flying Corps 86
Australian Government 85
Australian National Airways 86, 90
Australian War Memorial 207
Aviation Générale 143
Aviation Traders 177
Avro Anson 14

Avro Shackleton 73
Azores 16, 42, 133, 135

B

Bahrein 76, 208
Baker, Air Vice Marshal Brian 27
Baker, Mathew 187
Ball, Lt Henry Lidgbird, RN 99
Banak 70
Bantry Bay 128
Bardsey Island 17
Barrier Reef Airways 90, 162
Bathmat exercises 46
Battery, The, New York 125
Battle for Canton Island 79
Battle of Britain Flight 201
Battle of the Atlantic 12, 13, 16, 44, 65, 212
Baveystock 38
Baveystock, Flt Lt L H, DSO, DFC, DFM 29
Bay of Biscay 14, 15, 20, 30, 42, 52
Bay of Islands 119
Beachcomber VH-BRC 90, 94, 96, 115, 117, 119, 126, 156
Beaufighter 24, 25, 64, 66
Belfast 60, 94, 152
Belfast, HMS 149, 150
Belfast Lough 9
Bennett-Turner, Alan 173
Bergen 64, 69, 70
Bermuda 133, 134
Bermuda Yacht Club 136
Berre Lagoon 142
Biggin Hill 148, 167, 201
Billefjord 70
Björnebye, Capt S 64
Blackburn Aircraft Ltd. 178
Blair, Capt Charles F 120, 123, 124, 125, 128, 152, 194, 207
Blair, Capt Charles F (death of) 129
Blandford, Flt Eng Mark 133
Bleakley, Sgt R 78
Blenheim 20
Blind landing system 54
Blue Lagoon, guesthouse 107
BOAC 11, 70, 86, 93, 119, 120, 131, 158, 197, 209
Böhme, General Franz 64
Bomber Command, RAF 21, 23, 30, 51
Bora Bora 87
Bordeaux 148
Boscombe Down 202
Boston 125, 126
Boussiron 143, 144
Bowhill, Air Chief Marshall Sir Frederick 12, 14
Bowling Club, Lord Howe Island 106

Braby pontoon 77, 119
Brest 29, 37, 38, 43
Brindisi 209
Brisbane 90, 113, 208
Bristol Beaufighter 24, 64
Bristol Blenheim 20
Bristol Pegasus engine 51, 61
British Aerospace 183
British Aerospace Nimrod 148
British Airways 158
British Power Boat Company 197
British Virgin Islands 127
Brittany 41, 49
Bugge, Capt T 64
Bureau Véritas 145
Burgess, Capt J W 100, 112
Burgess, Mark 178, 186, 193, 197
Burns Philp South Pacific Trading Co. 99, 103
Bush, R W F 93

C

CAA (Civil Aviation Authority), UK 127, 145, 158, 166, 183, 191, 207
Cable and Wireless 79
Cabot Island 137
Cahill, Grp Capt C H, DFC, AFC 67
Cairns 77, 208
Cairo 208
Calcutta 208
Calf of Man 58
Calshot 11, 17, 18, 19, 20, 40, 44, 67, 71, 75, 76, 128, 138, 151, 163, 173, 193, 197, 206
Calshot Activities Centre 153
Calshot Spit 18, 211
Calshot Spit, lightship 152, 197
Canadair CL 215 143
Canberra, P&O liner 205
Canton Island 79, 80, 123
Caravelle 143
Caribbean 117
Caribou, De Havilland (Canada) 118
Carmarthen Bay 20
Carnegie, AVM DV, CB, CBE, AFC 78
Castle Archdale 20, 29, 44, 45, 54, 73
Catalina 15, 45, 61, 64, 73, 77, 86, 87, 88, 89, 90, 101, 102, 136, 158, 197, 210, 212
Centaurus, Empire Boat 100, 112
Certificate of Airworthiness 145, 153, 165, 171, 172, 174, 191, 189, 200, 203, 206, 207
Certificate of Fitness for Flight 158
Channel Islands 49, 165
Charlotte Amalie 126
Chartres, Frank 103, 105, 113, 114, 118
Chatham 199

Chatham Historic Dockyard   157, 164, 169, 172, 173, 197
Chatham Historic Dockyard Trust   169
Chatham Island   84
Chatham Islands   92, 181
Cherbourg   40, 148
Chichester, (Sir) Francis   100
Chieftain - motor launch   109
Chile   86
China Bay   77
Christiansted   126, 134
Christmas 1944, Pembroke Dock   47
Christmas Island   80
Civil Aviation Authority (UK)   127, 145, 158, 166, 183, 191, 207
Clyde   44
Coghlan, Michael   131, 139
Commonwealth Government (Australia)   111, 113, 115
Compass swing   187
Concorde   209
Coningsby, RAF base   201
Consolidated Catalina   15, 45, 61, 64, 73, 77, 86, 87, 88, 89, 90, 101, 102, 136, 158, 197, 210, 212
Consolidated Liberator   15
Convoy JW66   63
Cook Islands   78
Coral Route   92
Cork patrol   27, 29, 30, 34, 38, 40
Cowes   86, 200
Croix du Sud, French flying boat   143
Cromarty Firth   73
Crump, Sgt N   78
Cruz Bay   126
CSU electric heads   170
Cuckolds Point   149
Cullan, Brian   194, 197

D

Dan-Air   165, 168, 178
Dan-Air Engineering   155
Daniels, D T   173, 182, 183, 190, 191
Daniels, Joyce   174, 183
Danzig   43
Darwin   77, 100, 208
Daydream Island   90
DC3   168
DCA   (Department of Civil Aviation) Aus.   92, 95, 96, 172, 106, 111, 112, 113
De Havilland (Canada) Caribou   118
De Havilland Gypsy Moth   100
De Havilland Mosquito   43, 64
De Havilland Tiger Moth   20
De Havilland Vampire   143
Dept of Civil Aviation (DCA) Aus.   92, 95, 96, 106, 111, 112, 113, 172

Department of Transport (*See* DCA)   113
Derg Marine   194
Detwiller, Flt Lt L F   54, 55
Devonshire, HMS   66, 69
Dignam family   100
Dignam, Phil   114
Disarmament Wing, Allied, in Norway   71
DNL — Norwegian airline   71-2
Dokken, Capt Eric   204
Dönitz, Grossadmiral Karl   13, 14, 16, 26, 29, 34, 64, 65
Dorman, Jim   106, 107
Dornier flying boats   71
Douglas DC3   168
Down Island Cruises   126
Dubrovnik   209
Dundee   68
Dutch East Indies   100
Duxford Airfield   182

E

Eagle Mountain Lake   125
Easter Island   86
Easton, Flt Lt D A   20, 24, 25, 29, 30, 37, 38, 40, 41
Eisenhower, General Dwight D   27, 70
Electrical modifications   155
Electro Boat, U-boat   43
Emmott, Capt Ken   158, 163, 166, 171, 189, 193
Empire Boats   85, 94, 197
Engine failure, master rod bearing   171, 172
English Estates   157, 172
Entebbe Airport   181
Erne, Lough   45, 73
Essendon Airport   95
Etheridge, D H   152, 197
Eucumbene, Lake   119
Evans, Capt Andrew   146, 148
Evensen, Capt O G   63, 66, 70
Exactors   170
Excalibur III, Mustang   124
Excalibur VIII, N158J   123, 123, 126, 127, 128, 130, 135, 137, 141, 159
Export Certificate of Airworthiness   122

F

FAA   (Federal Aviation Administration) USA   122, 123, 126
Fabre, Henri   142
Faeroes   67
Fagan, Paul   134
Fairey Aircraft Company   200
Fajardo   126
Fanara   76
Fanning Island   79, 80
Fanua Lai   82

Far East Flight   85
Farnborough   189
Fastnet   40
Feathering engine   166
Federal Aviation Administration (USA)   122, 123, 126
Fenton, Stan   100, 112
Field Aviation South Africa   190, 191
Fiji   77, 90
Fiji, Governor of   77
Finnmark   69
Flanagan, Flt Eng Jim   125
Fleetlands, Naval Air Yard   201
Flight International   magazine   145
Flight Refuelling Ltd   152
Flying Boat, pub   152
Flying Boat Storage Unit   76
Flying Boat Test Flight   76
Flying Boat Union   12
Flying Fortress (Boeing B-17)   212
Fokker F27 Friendship   95, 111, 114, 115
Fornebu   69, 70
Foynes   128, 184, 194
French Navy   73, 144
French, Sqn Ldr W M DFC   59
Frigate Bird III   86, 87, 92
Froggatt, Flt Eng J D   87, 135, 137, 139, 141, 143, 146, 148
Fuel system modifications   155
Funafuti   79
Fysh, Sir W H, KBE, DFC   89

G

Galahad (Sir), RFA vessel   138
Galloway, Sgt A S   78
Gander   135, 137
Gander Lake   136
Gardiner, Capt Bluey   195
Garth's Voe   62
Gatwick   20, 189
Gauss, Flt Lt C M   46, 60
GEC Avionics   160, 180
Gee - radio nav. aid   49, 54
Gilbert and Ellice Islands   78
Giles, Flt Lt L E   46, 48, 49, 50, 52, 58
Gillies, Capt Ron N   91, 97, 120, 124, 125, 126, 128, 132, 138, 147, 148, 156, 174, 175, 194
Gironde Estuary   38
Goldthorpe, Flt Lt   56
Gouge, Sir Arthur   159
Grafton   87
Great Sound, Bermuda   135, 136
Greek Islands   209
Green, J   87
Greenland   69, 79

H

Grumman 115, 126
Grumman Goose 121, 129
Guinness Peat Aviation 184
Gunton, Squ Ldr M S 83
Gurnard, Isle of Wight 158
Gypsy Moth 100

H

Haakon, King of Norway 66, 69
Halifax 27, 29
Halton 165
Hamble Point 200
Hamilton Reach, Brisbane River 162
Hampshire County Council 157, 210
Hampshire, Wg Cdr J M, DFC 54
Hamworthy 86
Hand, Stephen 187
Handley Page Halifax 27
Happy Time 14
Haraldur, Norwegian sailing vessel 63
Harbottle, W J 183
Harland and Wolff 10
Harness, Capt (harbourmaster) 82
Harris, Sgt R C 30, 40, 42, 71
Harry, N A J, OBE 182
Hart, John 178, 183, 186
Hastings 160
Hawaii 79
Hawker Aircraft Ltd 165
Hawker Hurricane 45
Heads, The, Sydney Harbour 123
Heathrow Airport 189
Heavylift Cargo Airlines 133
Henry, Capt H 93
Henry VIII 18, 211
Herdla 70
Heron Island 90
High Tea 57, 59, 64
Himalaya, RMS 82
Hirtle, Plt Off F 59
Hitler, Adolf 13
Hobart 87, 90
Hobsonville 83, 84, 93
Hodgkinson, Sqn Ldr V A, DFC 102, 139
Holle, Noel 96, 126
Holman, A C 183
Holmedale, NZ coaster 84
Holmewood, NZ coaster 84
Holyhead 17
Hong Kong 80
Honolulu 124
Hookham, C R 183
Howea palm 99
Huffduff 16

Hughes, Howard 124
Hull plating 154
Hulton, Edward A S (Chs. 8-11)
Hurn 189
Hurricane (Hawker) 45
Hurricane Andrew 211
Hurricane, October 1987 174
Hythe 152, 197, 198
Hythe class flying boat 87, 158
Hythe Pier 198

I

Ile d'Yeu 38
Imperial Airways 79, 85, 100, 142, 145, 197, 208
Imperial War Museum 182
Inskip, M J D 154
Insurance policy 176
International Date Line 78
Ionian Isles 209
Ireland 136, 137
Irish television 196
Irvinestown 45
Isla Grande 127, 128
Islander (Ch. 7-11)
Isle of Wight 27, 158, 197, 199
Istanbul 209
Itchen, River 199

J

Janitrol heating system 95
Jeanne d'Arc, French cruiser 40, 80
Johns, Flt Sgt W H 35, 40
Jones, Robin 156, 165
Joyita 83
Joyita, S Pacific trading vessel 82
Juliet Flying Boats Inc 130, 145
Junkers 88 24, 25

K

Kabara Island 82
Kaldager, Cdr C R 67
Karachi 208
Kattegat 64
Kelly, C A 97
Kentia Palm 99
Killadeas 73
Killaloe 138, 194, 197
Kingsford-Smith, Sir Charles 86
Korangi Creek 76
Koro Sea 82
Krancke, Admiral Theodore 27, 30

L

Lagos 45
Laing, Flt Lt J A 78, 79, 82, 84
Lake Eucumbene 119
Lake Windermere (Festival) 201, 202
Land, Flt Eng John 158, 166
Land Rover 174, 176
Land's End 27
Lassiter, Fg Off R R 40
Last African Flying Boat 197
Latécoère 143
Lauthala Bay 77, 79, 81, 82, 83, 84
Le Pine, Wg Cdr R 79
Leanda Lei, guesthouse 105
Lee, D W 182
Lee-on-the-Solent 156
Leigh Light 22
Lepe 19
Lerwick 62
Lewis, Lester 105
Liberator 15, 29, 49, 170
Lillingston, Fg Off P J E 36
Limavady 57
Limerick 184
Lindeman Island 90
Line Islands 80
Lioré-et-Olivier 143
Lippincott, Harvey 171
Lisbon 133
Liverpool 202
Liverpool anti U-boat School 16
Llanion Barracks 48
Loch Ryan 11, 17
Lock, Albert 165, 178, 202
Lockheed Hudson 14
Logan Airport 125
London 150
London City Airport 189
Long Beach 124
Long Reach, River Medway 160, 161, 1890
Lord Howe Island (Ch. 7) 87, 90, 97, 99, 121
Lord Howe Island Board 113, 114
Lorentzen, S/Lt Ö 70
Lorient 14
Los Angeles 124
Lough Derg 137, 193, 196
Lough Erne 45, 73
Lower Pool 149
Luftwaffe 24, 29, 37, 52, 61
Lyme Bay 193
Lynch, P G (Benny) 135, 157

M

Mable, Capt Bill 133
MacFarlane, Ross 156, 159, 165, 178
Macfie, Plt Off D M 48
MacKenzie, Flt Lt K A 58
Maier, Flt Lt G P 49, 53, 54
Malabar 101
Malta 76
Mangeri, Joe 130
Manston 189
Marchwood Military Port 153
Mares, Flt Eng W H 145, 146, 148, 150, 155, 159
Marignane (Ch. 9) 139, 141, 145
Maritime Operational Conversion Unit 84
Marseille 134, 139, 145, 147, 148, 208
Martin Mars 158
Mascot 115, 119
Masterton, G C 189
Mathiesen, Capt Phillip 88
Maundrell, Capt L L 93, 97, 105, 115, 118, 123, 128
Mawer, Flt Sgt C C, DFM 40
McCready, Fg Off R A N 36
McDermott, Capt Don 133
McGoldrick Capt P J 196
McGrath, Fg Off R 82
Mechanics Bay 93
Medway City Council, Rochester upon 173, 177, 191, 210
Medway Estuary 160
Medway Heritage Foundry 180
Medway Ports Authority 161
Medway, River 157
Mermoz, Jean 143
Met Station, Lord Howe Island 106
Metox receiver 15, 21
Miami 210
Middlemiss, Capt S C, OBE 86, 90, 92, 93, 97, 109, 121, 122, 170
Miles, Bruce 208
Miles, Happy 130
Milford Haven 17, 24, 29, 38, 40, 59, 193
Mills, Flt Lt A D 46
Milorg, Norwegian resistance 67
Mistral 141, 143
Mokangai Island 77
Monkton, Capt B A W 86, 89, 115, 126, 133, 135, 139, 207
Monte Carlo 139, 145
Morgan Grenfell 20
Mosquito 43, 64, 66
Moulton, M F 157, 161, 163, 173
Mount Batten, Plymouth 20, 36, 38, 55, 61
Murdoch, K R 208
Murray, David 111, 112
Musée de l'Air (Paris) 92, 181

Museum of Transport and Technology 84
Musick, Capt E C 100

N

Nantes 148
Napier 84
Narbonne 148
Nauru 78
Nears, Graham 156, 165, 178, 193, 197, 203
Ned's Beach 109
Nelson, one-eyed pigeon 193, 197
Nesbitt, Fl Lt J G 49
Nesheim, Qm A 63
New Caledonia 78, 87
New Guinea 87
New Hebrides 78, 87
New York 126, 150
New Zealand 100
News Corporation (Australia) 208
Nichols, Mick 106
Nicholson, Flt Sgt G B 25
Niedermayer, Eric 183
Nimrod 148
No. 7 Covered Slip 157, 163, 165, 167, 174, 210
Noble-Campbell, Flt Lt K C 82
Noon, Alan 178, 183, 186
Norfolk Island 99, 100, 109
Normandy 27
North Haven 79
North Pole 124
North Sea 14, 66, 72
Northrop N3P-B floatplane 61
Norway, proposed visit 204
Norwegian Air Force 72
Norwegian Army 70
Norwegian Government 61, 66
Norwegian Navy 61, 68
Noumea 77, 87, 90
NSW State Government 111, 113

O

Oban 61
Ocean Island 78
Ocean View, guesthouse 99
O'Dell, Flt Sgt N C 49
O'Hara, Maureen 121, 124, 128, 132, 134, 159, 184
Olav, Crown Prince, of Norway 66
Open Boat Slip 172
Operation Cork 30, 36, 37
Operation Overlord 27
Operations Manual 168
Orient Airways (Aviation Heritage) Limited 209, 210
Orient Express 209

Orsova, SS (P&O-Orient Lines) 109
O'Shaughnessy, Margaret 177, 184, 195, 197
Oslo 68, 71
Osprey, HMS (Shore establishment) 12

P

P & O-Orient Lines 109
Pacific Chieftain VH-BRE 90, 91, 94, 95, 96, 106, 107
Pago Pago 100, 123
Pan Am 120, 134
Pan American Airways System 79, 100
Pan American Clipper 123, 184
Park, Fg Off D A P 49, 58, 59
Patterson, Fg Off J 78
Pearl Harbour 124
Pegasus engine 51
Pembroke Dock 17, 18, 20, 29, 37, 38, 39, 40, 41, 42, 44, 45, 67, 85, 193
Pembrokeshire 55
Perry, Flt Lt James 27
Phillip, Capt Arthur, RN 99
Phillipsburg 126
Phoenix 19
Phoney War 13
Pinetrees, guesthouse 99
Pitt Island 84
PLA (Port of London Authority) 147, 151
Plymouth 20
Plymouth Sound 36
Ponta Delgado 135
Pool of London 30
Poole, Dorset 86, 128
Poole, Flt Lt Aubrey 17, 19
Port Headland 85
Port Jackson 99
Port of London Authority (PLA) 147, 151
Potomac River 125
Poulson, Capt Christian 90
Pratt & Whitney R1830 engine 190
Pratt and Whitney 170
Pratt and Whitney Twin Wasp 61
Prien, Kapitänleutnant Günther 14
Princess flying boat 143
Pringle, Lt Gen Sir Steuart 169, 173
Propellers 204
Provence 40, 141
Puerto Rico 126, 129
Purvis, Capt G H 87

Q

Qantas 85, 87, 89, 119, 208
QEC 167
Queen Mary, ex-Cunard liner 124

Queen's Island, Belfast 9, 61, 73
Quisling, Vidkun 68

R

RAAF 85, 91, 93, 118, 170
RAF High Speed Flight 152
RAF Museum, Hendon 181
Raffles 81
RAI, French airline 92
Rangoon 208
Rathmines 87, 102, 207
Rathmines, RAAF base 101
Red Army 67, 69
Redland Bay 113
Reef Hotel, St Croix 133
Rennes 148
Resorts International 130
RFC (Royal Flying Corps) 86
Richardson, Fg Off N E 78
Richmond, Sgt R R 72
Ringe, S/Lt T 63
River Itchen 199
River Medway 157
River Shannon 128, 135, 137, 194
River Thames 146, 151
RNAF 204
RNZAF (Ch. 5) 73, 92, 110, 166, 181
Road Town, Tortola 126
Roberts, Alwyn 167, 173
Roberts, Capt Gilbert, RN 15
Robinson, Flt Sgt R 42
Rochester 51, 61, 85
Rochester Airport 160, 180
Rochester upon Medway City Council 173, 177, 191, 210
Rose Bay 85, 86, 90, 93, 95, 102, 103, 109,
        110, 111, 113, 115, 119, 120, 121, 123
Rosyth 66
Royal Aero Club 12
Royal Aeronautical Society, Medway branch 157, 165, 173
Royal Canadian Navy 40
Royal Engineers 182
Royal Marines 99
Royal Marines Cadets 169, 178
Royal Naval Air Service 17
Royal New Zealand Air Force 73, 92, 110, 166, 181
Royal Norwegian Air Force 204
Royal Oak, HMS 14
Russian convoys 45, 63
Ruth, Squ Ldr W D B, DFC 30
Ryanair 184, 187, 197

S

SAAF 73
Saltus Bay 136
Saltus Island 135
Samoa 123
Samoan Clipper 100
San Francisco 100
San Juan 126
Sandringham 73, 86, 87, 94, 96, 111,
        115, 120, 121, 126, 127, 162, 170
Sandstrom, Doug 131
Sapeurs-Pompiers 146
Saro London 14, 17
Satapuala Bay 82, 123
Saunders-Roe 86, 159, 200
Saunders-Roe Princess 143
Scapa Flow 14
Schneider Trophy 152
Schnorkel U-boats 26, 29, 36, 37, 43, 46, 57
Science Museum 156
Scilly Isles 20, 34, 54, 36
Scottish Aviation 131
Seaford 148
Seaplane Works 51, 159, 160, 165, 174, 182, 202
Searle, M C 171, 189, 197, 198
Sécurité Civile 143
Seletar 76
Shackleton 73
Shadwell Basin 149
Shannon, River 128, 135, 137, 194
Shanwick Control 137
Shelton, Les 127
Shetland Islands 62, 64
Shipway, C F 172
Short and Harland Ltd. 9
Shorts Belfast 133
Short, June 174
Short Sandringham 72
Short Singapore biplane flying boat 77
Short Slip (newsletter) 50, 59
Short Solent flying boat 84, 126, 158, 197, 210
Short Stirling 22, 161
Sikorsky 123
Sikorsky S42 100
Sikorsky VS44 120
Singapore 80, 81, 85, 208
Sissons, Capt Keith 133, 134
Skins, fitting of 179
Skysport Engineering 168
Sliwinsky, Felix 144
Smith, Fg Off R L 49
Smith Lake 119
Snowy Mountains 119
Soerabaja 77

Sola 68, 70, 71, 204
Solent, The 18
Solent flying boat 84, 92, 126, 158, 197, 210
Solomon Islands 87
Sono-buoy 57, 75
Sotheby's 204
Sourabaya 208
South African Air Force 73
South Atlantic Fund 146, 151
South Pacific Air Lines 126
Southall, Plt Off I F 38, 62
Southampton 18, 85, 119, 135, 198, 203
Southampton Docks 197
Southampton Hall of Aviation 181
Southampton Water 18, 19, 40, 71, 152, 197
Southern Cross, N158C 126, 127, 128, 132,
        138, 141, 152, 194
Soviet Air Force 45
Spirit of Foynes 187, 193
Spumedrift, yacht 112
St Ann's Head 40
St Croix 125, 126, 129
St George's Channel 193
St John 126
St Maarten 126
St Mary's Sound, Scilly Isles 37
St Nazaire 38, 148
St Pierre Point 136
St Thomas 126, 129, 134
Start Point 40, 49
Statue of Liberty 125
Stavanger 68, 69, 72
Stevenson, Robert Louis 81
Stewart, Flt Lt D B, (MD) 47, 50, 55
Stirling 22
Stirling bomber 161
Stord, HNMS 66
Straddle, Plt Off (mascot) 55, 56
Street, J A 155
Strumble Head 17, 193
Sud Aviation Caravelle 143
Sullom Voe 17, 44, 62, 67, 68
Sumner, Wg Cdr J R, DFC 47, 48, 50
Sunderland Flying Boat Foundation 206, 209
Sunderland Limited (Chs. 8-11)
Sunderland Type Record 183
Supermarine Southampton flying boat 85
Supermarine Stranraer 14
Supply, HMS 99
Surrender — European theatre 64
Sydney 85, 90, 99, 100, 109, 123, 208
Sydney Harbour 85, 97
Sydney press 85
Sydney Water Airport 119

T

TAA 90
Tahiti 78, 86, 87
TAI, French airline 87
Tamiami Airport 211
Tarawa 78, 84
Tasman Empire Airways Limited 84, 90, 92, 93, 119
Tasman Sea 85, 86, 119
Taylor, Brian 184, 187
Taylor, Sir Gordon 86, 92, 100, 135
Te Whanga Lagoon 84
Teague Bay 133
TEAL 84, 90, 92, 93, 119
Tenyo Maru, Japanese fishing vessel 82
Thames, River 146, 151
Third Reich 65
Thorne, Gen Sir Andrew, KCB, CMG, DSO 66, 68, 69
Thunderbolt Pier 169
Tiger Moth 20
Tirpitz 69
TOA (Trans Oceanic Airways) 87, 89, 90, 93, 119, 135
Tokelau Islands 81, 82
Tompkins, Flt Lt E J E 76, 77
Torgersen, Flt Lt H 80, 82
Tortola, BVI 126
Toulouse 148
Tower Bridge 30, 149, 150
Tower of London 149
Trans Australian Airways (TAA) 90
Trans Oceanic Airways (TOA) 87, 89, 90, 93, 119, 135
Treaty of Versailles 13
Trippe, Juan 79
Tromsø 69, 70
Trondheim 69, 71
Tulagi, MV (Burns Philip) 109

U

U-107 38
U-385 38
U-621 34
U-2511 64
U-3501 43
U-boat, type VIIC 26
U-boat, type XXI 43, 64
U-boat, type XXVI 43
Ulm, Capt C 86
United States 16
United States Air Force 124
United Technologies Inc. 171
US Army VIII Corps 43
US Navy 73, 135
Ushant 27

V

Vampire 143
Vavau 82
Velox, John 127
Venice 209
Vickers, Flt Lt F J 78, 79
Vickers Wellington 22
Victory, HMS 193
Viking, HM Submarine 27
Virgin Gorda 133
Virgin Islands 125
Vold, S/Lt 69
VTS Southampton 205

W

Waitemata Harbour, Auckland 84, 100
Walker, Capt F J, RN 38
Wallis Island 83
Wallis Lake 119
Walter turbine 43
Walters, Flt Lt I F B, DFC 37
Wapping Pier 149
Ware, Wg Co 'Mo' 135
Washington 125
Watlington, Capt Hugh 136
Weaver, Sqn Ldr F H Q, AFC 76
Weeks Air Museum 210
Weeks, Kermit 210
Weir, Flt Sgt W J 41
Wellington 22, 29
West End 126
West Malling 198
Western Samoa 81, 82
Westland 200
Westland Sea King 72
Westland Wessex 148
Whittome, Wg Cdr R W 67
Wig Bay 11, 16, 17, 18, 73, 76
Wilcher, Capt Bill 97, 110
Wilson, Clive 114
Wilson, Roy 105, 114
Windermere, Lake 201, 202
Windy Point 106
Woodhaven 68
Woolett, Bob 165, 174, 176, 178, 193
Woolnough, Harry, OAM 112
World Heritage Listing 118
Wrathall, Capt Gary 166

Y

Yasawas island group 79
Young, Capt Reg 158

Z

Zodiacal Lights 82

Numerals

5 Squadron, RNZAF (Ch. 6) 77
6 Squadron, RNZAF 83
10 Squadron RAAF 20, 61, 85, 206
17 Port Regiment 153
18 Group, Coastal Command 61
19 Group, Coastal Command 20, 29, 34, 37, 42, 46, 49, 56
20th Mountain Army (Germany) 67
24 Field Squadron Royal Engineers 163, 182
57 MU, RAF Maintenance Unit 11, 16, 73
201 Squadron RAF (Ch. 2) 17, 45, 79, 148
210 Squadron, RAF 64
228 Squadron, RAF 29
272 MU, RAF Maintenance Unit 73
330 (Norwegian) Squadron RAF (Ch. 4) 61, 68
330 Squadron, RNAF 204
333 (Norwegian) Squadron RAF 68
422 Squadron RCAF (Ch. 3) 45, 61, 67
423 Squadron RCAF 45
461 Squadron RAAF 38, 53, 67

Baff, K C — *Maritime is Number Ten*. Baff, 1983.

Blair, Charles — *Red Ball in the Sky*. Jarrolds, 1970.

Bowyer, Chaz — *Men of Coastal Command 1939-45*. William Kimber, 1985.

Chichester, Francis — *The Lonely Sea and the Sky*. Hodder and Stoughton, 1964.

Compton-Hall, Richard — *The Underwater War 1939-45*. Blandford Press, 1982.

Driscoll, Ian — *Flightpath South Pacific*. Whitcombe and Tombs, 1972.

Franks, Norman — *Search and Kill*. Aston Publications, 1990.

Fysh, Sir Hudson — *Wings to the World*. Angus and Robertson, 1970.

Gunn, John — *Challenging Horizons*. University of Queensland Press, 1987.

Gunn, John — *The Defeat of Distance*. University of Queensland Press, 1985.

Hendrie, Andrew — *Flying Cats*. Airlife, 1988.

Herington, J — *Air Power Over Europe*. AWM, 1963.

Jablonski, Edward — *Seawings*. Robert Hale, 1974.

Lindsay, Donald — *Forgotten General — A Life of Andrew Thorne*. Russell, 1987.

Livock, G E — *To the Ends of the Air*. Imperial War Museum, 1973.

Mordal, Jacques — *The French Navy in World War Two*. US Naval Institute, 1959.

Moss, M & Hume, J R — *Shipbuilders to the World*. Blackstaff Press, 1986.

Murley, Clare & Fred — *Waterside — A Pictorial Past*. Ensign Publications, 1991.

National Archives of New Zealand — *Air 143/5* and *Air 144/3*.

National Transportation Safety Board, Washington DC. — *Report NTSB-AAR-79-9*.

Price, Alfred — *Aircraft Versus Submarine*. Janes, 1980.

Rance, Adrian — *Fast Boats and Flying Boats*. Ensign Publications, 1989.

Rawlings, John — *Coastal, Support, and Special Squadrons of the RAF*. Janes, 1982.

Southall, Ivan — *They Shall Not Pass Unseen*. Angus and Robertson, 1956.

Stewart Middlemiss Story, The — Aviation Heritage. Vol. 25 Nos. 1 & 2.

Taylor, Sir Gordon — *Bird of the Islands*. Cassell Australia, 1964.

Thorne, General Sir Andrew — Typewritten report, Imperial War Museum, London.

Whinney, Bob — *The U-Boat Peril*. Arrow, 1989.

Williams, Mark — *Captain Gilbert Roberts, RN*. Cassell, 1979.

Wilson, M & Robinson, A — *Coastal Command Leads the Invasion*. Jarrolds, 1945.

For a complete list of our books write to us at:
Ensign Publications
2 Redcar Street
SOUTHAMPTON
UNITED KINGDOM
SO1 5LL

FAST BOATS
and
Flying Boats

A Biography of Hubert Scott-Paine

On the 20th September 1916, Hubert Scott-Paine became the sole owner of the Supermarine Aviation Company Ltd, and after just seven years was able to sell his company for £200,000.

In the following years he set about revolutionising the motor boat industry. The vast portion of this book deals with the fascinating and previously almost unknown story of the British Power Boat Company.

Scott-Paine started the company and drove it forward to become the most successful developer and manufacturer of fast air sea rescue craft and motor gun boats of World War II.

At Supermarine and at British Power Boats he first secured and then nurtured the creative talents of two of the country's leading designers in their respective fields — R. J. Mitchell at Woolston and George Selman at the boat 'factory' at Hythe in Hampshire. Mitchell went on to design the Spitfire, and Selman's plans became the basis for the legendary American PT boats of World War II.

"... a highly entertaining book which will also be well received by serious students of maritime and aviation history. Compulsive reading."

Motor Boat and Yachting

*192 pages · 90 illustrations
Appendices · Index
Hardback*
£14.95 (plus £1.50 p+p)

Sunderland "Isla[nder]"
CABIN ARRANGEMENT, AS F[...]

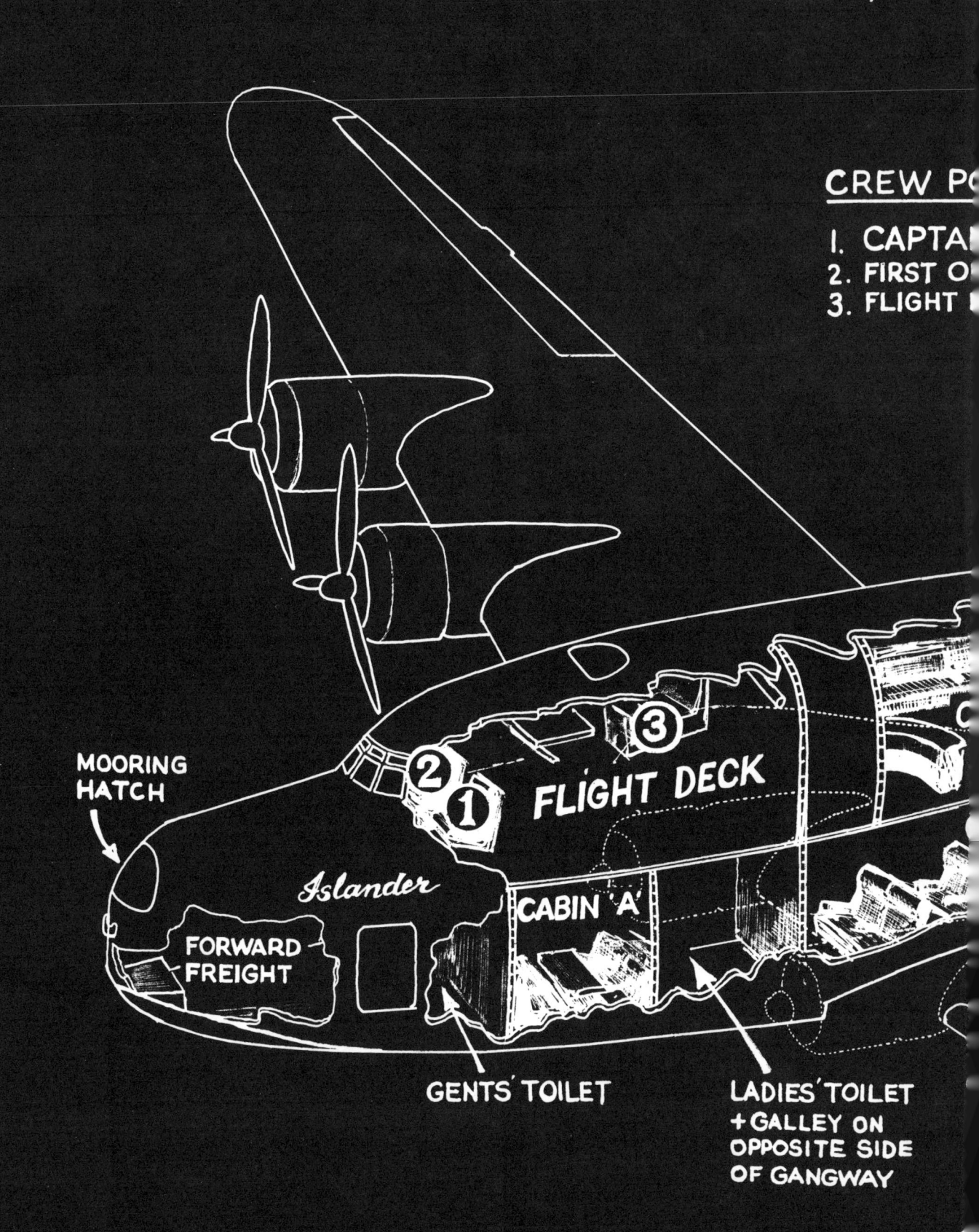

CREW P[OSITIONS]

1. CAPTA[IN]
2. FIRST O[FFICER]
3. FLIGHT [...]

- MOORING HATCH
- FLIGHT DECK
- *Islander*
- CABIN 'A'
- FORWARD FREIGHT
- GENTS' TOILET
- LADIES' TOILET + GALLEY ON OPPOSITE SIDE OF GANGWAY